Language and learning in the teaching of geography

Edited by
Frances Slater

ROUTLEDGE
London and New York

First published 1989
by Routledge
11 New Fetter Lane, London EC4P 4EE
29 West 35th Street, New York, NY 10001

© 1989 Frances Slater

Wordprocessed by Amy Boyle Wordprocessing
Printed and bound in Great Britain by Mackays of Chatham

All rights reserved. No part of this book may be reprinted or reproduced or utilized in any form or by any electronic, mechanical, or other means, now known or hereafter invented, including photocopying and recording, or in any information storage or retrieval system, without permission in writing from the publishers.

British Library Cataloguing in Publication Data

Language and learning in the teaching of geography.
   1. Great Britain. Schools. Curriculum subjects.
   Geography. Teaching
   I. Slater, Frances
   910'.7'1041
   ISBN 0-415-03213-X

Library of Congress Cataloging in Publication Data
   0-415-03213-X

CONTENTS

List of contributors
Acknowledgements
Introduction   1

Part I Language and learning

1 Language and learning
   Frances Slater   11
2 Writing in a humanities classroom
   Daniel Lewis   39
3 A case study of language and learning in
   physical geography
   Sharon Hamilton-Wieler   59
4 Talking and problem-solving: Reality
   Orientated Problem Solving questions:
   a strategy for inducing productive
   verbalization for externalizing and
   correcting misconceptions in map work in
   secondary schools
   Julie I.N. Okpala   71
5 Language and learning in multicultural
   education
   Daniel Lewis   92

Part II Language and concept development

6 Concept maps and children's thinking: a
   constructivist approach
   Anthony L. Ghaye and Elizabeth G. Robinson   115

Contents

7 The feeling tone of childhood: children writing about their environment
Ann Jones   141

Part III Language and ideology

8 Language and ideology in geography teaching
Rob Gilbert   151
9 The ideology of geographical language
Richard Henley   162

Part IV Language and reflection

10 'A teacher, an adult or a friend?'
Anthony L. Ghaye   175
11 Writing as reflection
Margaret Roberts   193

Index   212

CONTRIBUTORS

Anthony L. Ghaye is Senior Lecturer in Professional Studies at Worcester College of Higher Education. He has published a number of papers and articles in the field of children's learning.

Rob Gilbert is Dean of Education in the James Cook University of North Queensland. He is the author of *The Impotent Image*.

Sharon Hamilton-Wieler is Assistant Professor of English in the School of Liberal Arts at Indiana University in Indianapolis. She has published articles on writing across the curriculum, theories of language and learning, and the politics of literacy in the United States, Canada, and England.

Richard Henley is Head of Geography at Acland Burghley School, north London. He completed an M.A. in the University of London Institute of Education and his chapter reports research undertaken for that degree.

Ann Jones is a geography teacher at Chesterton Community College in Cambridge. She is continuing research begun at University College, London, into children's responses to environments.

Daniel Lewis is an advisory teacher in the London borough of Haringey. He has written in *Teaching Geography* on language and geography teaching.

Julie I. N. Okpala is Lecturer in Geography Curriculum in the Department of Education in the University of Nigeria, Nsukka. She has contributed several articles to research publications of the International Geographical Union on geography education.

Contributors

Elizabeth G. Robinson was formerly Head of Geography in Feltham Community School, west London. She completed an M.A. at the University of London Institute of Education and her joint chapter reports research undertaken for that degree.

Margaret Roberts is Lecturer in Education (Geography) at the University of Sheffield. She has just started working on a four-year research project investigating the introduction of a National Curriculum in Geography in secondary schools.

Carole Robson teaches English Language and Literature at Greenhill Tertiary College in Harrow. At the College she is also involved in a Language Development Group which is trying to find the best ways of tackling Language Across the Curriculum.

Frances Slater is Senior Lecturer in the geography section of the University of London Institute of Education. She is the author of *Learning through Geography* and other publications and articles in geography education and teacher education.

ACKNOWLEDGEMENTS

I should like to acknowledge the receipt of a small grant from the International Geographical Union to defray the costs involved in the preparation of the typescript of this book.

INTRODUCTION

OUR PURPOSES

The relationship of language and learning in teaching geography is not one of the most frequently contemplated or researched areas in geography education. Those wishing to do research are more likely to have an interest, for example, in aspects of map reading and comprehension, children's spatial abilities, fieldwork, the problems and opportunities of mixed-ability teaching, computer-assisted learning, some aspect of curriculum development such as environmental education, the history or evaluation of a curriculum project, values education and the teaching of controversial issues. The list could be longer but it rarely includes considerations of language and learning.

The purposes of this collection of essays are several. First, the collection demonstrates that there are a number of people working in geography education and language generally who attribute significance to the relationships between language and learning and have explored them in their own ways in geography teaching. Standing together, the essays also serve to make tentative statements about this area of learning in geography. These statements, it is hoped, will motivate others to take the work further.

The collection also illustrates a breadth and diversity of interest and begins to show what a large area for research we have when we talk about learning in geography and the significance and role of language in the process. It is also important to realize that the essays offer a kaleidoscopic view of research techniques available for investigating aspects of language and learning. These range from the strictly scientific to the more subjective. For this dimension, too, the volume is valuable, and each essay gives insights into the strengths of the selected methodologies. Contrasts of methodology illuminate.

Introduction

LANGUAGE AND LEARNING

There is probably one thread which most strongly runs through the essays: acceptance of the ideas of the Role of Language in Learning movement sometimes known as the Language across the Curriculum school, developed in the 1970s, carried and evolved into the 1980s and perhaps first signalled in Barnes *et al.* (1969) and Britton (1970).

In this book, I think, the threads of ideas from the Role of Language in Learning movement help us to make sense of essays which otherwise may seem disparate in focus and style. Part I focuses on the theory which may inform language practice in classrooms and the role of language in learning as demonstrated in student reading, writing, or talking.

In my introductory chapter I attempt to record how I first met ideas relating to language and learning; how I have related them to other work; and how I have interpreted them for my own work with geography graduates following the Postgraduate Certificate of Education (PGCE) course. The chapter is both a personal account and then a report of how other geography tutors working for the most part in university departments of education view and interpret the ideas of Barnes, Britton, Rosen, and Martin today. Finally I describe research work by Carol Robson, central to geography teaching, which examines the relationship between a sixth-form student trying to learn geography, the text she is trying to read, and the role the teacher has or has not played as mediator between the text and student.

In the second chapter Daniel Lewis writes about two bilingual pupils of different levels of achievement in the context of a humanities classroom. Coming from a Language across the Curriculum perspective, Lewis provides particular observations of Fatima and Ferdinand and their writing, and moves on to use existing research to discuss the effect of the classroom environment on the writing process; the ability of the teacher to recognize individual writing needs; and the influence of personal and social factors on their writing. In this way, we are appraised of some of the issues, relevant literature, and practical methods of fostering writing in bilingual pupils. This chapter, focused specifically on two pupils, sets the scene for the wider multicultural matters which Lewis takes up in chapter 5.

Following chapters 1 and 2, Sharon Hamilton-Wieler needs to spend less time on setting out the language and learning framework she is operating in. All three chapters share a common framework. In a case study style of research she response to, coming to terms with building language into learning and learning into language. Evidence is presented anecdotally. Clues to significant learning are found in Nigel's (the student's) writing, clues which point to the crucial role of personal experience and a sense of discovery in learning. When these two variables met was Nigel at last and at that point writing for himself? He had shifted audience - valuably.

As a piece of research Julie Okpala's work stands in contrast to Hamilton-Wieler's. It is a scientific undertaking with an objective style and language to match. She investigates a problem with its related hypotheses and sets to work with definitions and ways of classifying and statistically testing outcomes. The importance of her work in this volume lies in the different language outcomes she found to be characteristic of different ways of trying to teach map work. In the realistic problem-solving approach she devises, the opportunity pupils have to talk to teachers and other pupils reveals misconceptions and helps to clear them up. Dr Okpala relates such results to Language across the Curriculum notions that expressive (exploratory) talk facilitates the clarification of ideas. Although differing greatly in style and research methodology it is interesting that both this and the preceding chapter find it important that students have a sense of problem, while in Carol Robson's work, described in the first chapter, it was almost that the student had not been given enough sense of a problem to help differentiate out meaning.

The final chapter in Part I looks at language and learning not so much in relation to research into individual pupil reading, writing, or talking as in terms of how best to promote these activities in the context of the multicultural classroom. Daniel Lewis signals the conflicting and overlapping ideologies in the field, then seeks to show how policies of (1) validating the home language of students and (2) yet being aware of using language to challenge and change racist attitudes and behaviour relate to four concerns. Lewis essentially reviews work and opinion and points to good practice in the areas of dialect and bilingualism; general language strategies; specific language strategies; and language to challenge racism. This chapter relates to others in the collection in that good practice links with the language and learning ideas discussed in previous chapters, and his attention to the language of racism looks forward to Gilbert's and Henley's chapters on language and ideology and the language of the social sciences.

Introduction

LANGUAGE AND CONCEPT DEVELOPMENT

The chapters in Part II illustrate the significance and role of language and learning from a different angle. They reinforce our awareness that concept development is partly mirrored in language. Ghaye and Robinson, using a concept mapping technique as a research tool with classes they have taught - the teacher as researcher - try to understand how children comprehend and structure concepts in ways that make sense to them. The technique may be seen as one to be added to the repertoire of language strategies in the classroom. The concept map gives opportunities to sort out and clarify ideas and link ideas together. It gives us a view of learning in transition. As the authors suggest, concept mapping can also be undertaken by groups of children. Indeed, informed by the Barnes *et al.* school, the work points to the value of changing the audience. Children were asked to prepare concept maps for their teacher. They might also be asked to prepare them for themselves, for someone else, and so on.

In Ann Jones's chapter we get a glimpse in children's essays, in contrast to their concept maps, of their concepts of their environments. The children write of many facets of their environments as explored, known, and felt. Children's subjective essays form the data of a humanistic piece of research, i.e. a piece of research where feelings are not to be excluded. There is a message here. Some claim that too much of the writing, which is writing for geography in school, leaves out the feeling component. Awareness of this has been growing, I think, and such writing belongs within a Barnes-Britton-Rosen-Martin framework. All suggest we should widen the range of writing opportunities to bring in the children's world. Jones finds that children write with 'fluency, immediacy, detail, and freshness'. There is no communication gap here between their experience and what they write of.

LANGUAGE AND IDEOLOGY

From the children and their concepts we move in Part III to language and the concepts it embodies. As I understand it, people like Burgess (1984), who were also part of the London language team in the 1970s, have moved on to concern about ideology and language in the 1980s. This development has also taken place among geographers. Rob Gilbert argues that although the social sciences, of all subjects, should give students a critical insight into their society, they do not achieve any such goal. The view, the representation given of society in texts and schools is a restricting one. Gilbert's analysis of the language of some geography texts supports such a thesis. Through abstract, detached language, social forces are made general, impersonal, and natural, not personal, immediate, and controllable. Language is loaded, not transparent. Where Lewis wishes to challenge racism, Gilbert wants to challenge ideologies, and *status quo* ideologies in particular. Both want students to realize the loaded nature of language and to use language to challenge language.

The teacher as upholder of critical thinking needs to be conscious of this relationship between language, ideology and what is learned. It can seem obvious to state that language carries our cultural experience. We need to ask: whose cultural experience? And who benefits from a particular view? Henley, in his chapter, continues to elaborate on a view of language which does not see it as neutral or value-free. He examines the legacy of quantification and positivism on geography's language and illustrates the metaphorical impact of this. Natural science metaphors seem to dehumanize and objectify people, society, and social processes. If scientific geography masks, he is not entirely happy with humanistic stances either, and makes a plea for classroom teaching which examines all discourse. The achievement of such a goal would give a new meaning to the phrase 'language and learning'. From an examination of the language-loaded nature of racism and sexism perhaps we shall move more strongly in the next decade into an examination of language used to describe society and the role this plays in learning about society.

LANGUAGE AND REFLECTION

The final section is closer to earlier themes on the role of language in learning in its exploration of children's writing. In addition, from a research point of view the two chapters illustrate the practice and value of diary-keeping and demonstrate the effects of writing for oneself. Writing a diary, as a way of talking to oneself and the teacher, is one strategy for shifting from a teacher-centred to a pupil-centred stance and for showing up what children are learning and thinking about. As Ghaye observes, pupil diaries can help to develop more shared realities, realities enhanced by establishing and maintaining a dialogue between teacher and pupil through the diaries. Through diary-keeping, Ghaye also stresses, the relationship side of teaching and learning can be built up. The

Introduction

richness of the examples given by Anthony Ghaye and Margaret Roberts testifies eloquently to the value of diary-keeping for achieving reflection and a reordering of understanding.

Margaret Roberts's experiments with learning logs (diaries) extend from PGCE students writing about their methods course, to sixth-form students keeping a log during fieldwork, to sixth-formers making diary entries for their teacher about a new geography course. The experiments support many of the contentions of the Barnes *et al.* school. For example, diary entries are seen to have a role in reconstructing knowledge, clarifying thinking, and, for the teacher, understanding the process the writers are going through. Roberts pays particular attention to, and points to the significance of, how the diary-keeping exercise is set up. How diaries are introduced and responded to is crucial to their success in promoting learning and reflection.

I believe in the value of the ideas which inform and link the chapters of this book. Not long ago, in conversation about language and learning, language and talking, an educator said deprecatingly of the role of language ideas, 'Oh, kids can talk about this liquid being poured into this chemical and it fizzles up, but at some stage they have to be able to use terms like "chemical reaction" and so on. It's this necessary knowledge of terms and new concepts which the language team have never accepted.' I was bound to reply that I saw that as a caricature, a very partial understanding of the Barnes *et al.* work; that in fact talking to learn was seen to be part of the process of coming to an understanding of concepts and terms recently and newly encountered, of fitting old concepts like 'fizzling' to new ones like chemical reaction.

I hope that this collection fails to spawn any further misinterpretations and misconceptions of the view of the role of language in learning as Barnes, Britton, Rosen, and Martin developed it. Indeed, I hope that we have succeeded in presenting evidence and results of value to teachers who have a responsibility both for children's learning and understanding and for mediating such learning through their own language and the children's languages. The role of language in learning, as contributors to this book see it, is finally all about the process of learning and facilitating learning. This is the very point which the person quoted above has failed to hear, for whatever reason, though, ironically, a colleague of some of those central workers in the Language across the Curriculum movement over many years. Communication gaps are ever with us!

REFERENCES

Barnes, D., Britton, J., and Rosen, H. (1969) *Language, the Learner and the School*, Harmondsworth: Penguin.

Britton, J. (1970) *Language and Learning*, London: Allen Lane.

Burgess, T. (1984) 'The question of English', in M. Meek and J. Miller (eds.), *Changing English*, London: Heinemann.

Part I

LANGUAGE AND LEARNING

Chapter One

LANGUAGE AND LEARNING

Frances Slater

LANGUAGE ACROSS THE CURRICULUM

One of my very first in-service tasks after moving from teaching to teacher education was to lead a group discussion on what had become known as the Language across the Curriculum research and recommendations. Nancy Martin had given an introductory lecture in one of the dull and grimy lecture halls of the University of London Institute of Education, then located in Senate House. The year was, I think, 1974.

Her talk and the general message interested me greatly. The relationship of language to learning was not something I had ever been led to or had time to think about as a teacher. I suppose I assumed that it was partly through language that one learned; that was common sense.

After the in-service day I tried to find out more of the message. It seemed most often to be understood in the terms that every teacher is a teacher of English, every teacher has a duty to develop children's reading and writing abilities. Very few people understood it as a distinction between language as communicating, conveying meaning, and language as part of the activity of learning. The 1979 DES publication *Aspects of Secondary Education* was one of few to put it so clearly.

Some geography teachers had never heard of Language across the Curriculum. Others openly admitted they did not understand. Others were dismissive of it, sometimes as a kind of play-way method of learning, letting kids talk. Others thought it was soft on making mistakes: anything was correct. There was ignorance, hostility, and misunderstanding. This too puzzled me, since a reading of the project book, Martin *et al.* (1976), though not revealing to any great extent the texts behind their text, did imply in the report of their research a fairly simple formula which any teacher in any subject might employ for encouraging and opening up writing and speaking (language) activities in their classroom. I have summarized my understanding of this elsewhere (Slater, 1982) and shall briefly repeat that it seemed to be a matter of 'Vary the audience, vary the function', thus:

1. Student to self, as in 'A diary of my journey to .../the visit to ...'.
2. Student to trusted adult, for example, 'What I learned from the film/the field work at ...; What I see on this map/in this photograph ...'.
3. Student to student as partners in a dialogue, for example, discussing a particular concept, participating in a game.
4. Student to teacher in rough drafts of field reports or essays where comments are made to assist, diagnose strengths and weaknesses, suggest improvements before a final draft is marked for examination or test purposes. (From a rough to final draft represents a move from expressive to transactional which is also true of 5 below.)
5. Student to student, writing when position papers of speeches are being prepared. Alternatives 3 and 4 also provide opportunities for collaborative writing.
6. Writer to readers (relatively unknown audiences), including the design of advertisements, preparation of letters to newspapers or town planners. In such tasks language is likely to move from the expressive to the transactional.

THE ROLE OF THE AUDIENCE

The idea of varying the audience linked for me with Piaget's work, which I knew from *The Language and Thought of the Child* (1976), where his interpretations of experimental work led him to conclude that developing a sense of another person as audience is an essential variable in the child's development from egocentric to socialized speech. Vygotsky, on the other hand, points to audience also but sees language development differentiating into two functions from a social, not an egocentric, base. The communicative function of language (speech), Vygotsky claims, *arises from social contact*, and he argues that the earliest speech of the child is social, not egocentric only. 'At a certain age the social speech of the child is quite sharply divided into egocentric and communicative speech ...' and 'egocentric speech, splintered off from general social speech, in time leads to inner speech, which serves both autistic and logical thinking' (1962:19).

Figure 1.1

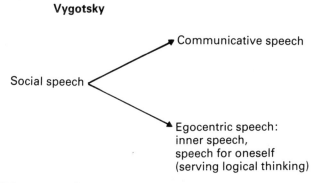

Diagrammatically, the difference between Piaget and Vygotsky may be represented as in Figure 1.1. There are other theoretical ideas which, together with those of Piaget and Vygotsky, provide a background to the meanings which the London Writing Research Group/Language across the Curriculum Group gave to the notions of expressive (better called exploratory, in my opinion), transactional, and poetic talk and writing. These concepts of expressive, transactional, and poetic grow out of ideas of speech functioning, on the one hand, egocentrically and, on the other, for communication. The sense of the role of audience, together with the functional distinctions between expressive, transactional, and poetic forms of speaking and writing can be seen to be grounded in Piaget and Vygotsky as well as in others such as D.W. Harding and M.K. Halliday. Indeed, in the elaborations which Vygotsky may be seen to develop from Piaget's work, expressive talking or writing (coming from inner speech) is seen to sit very firmly within the notion that thinking develops out of going over thoughts and ideas: rubbing them together, from which process they emerge refined, ready to be further rotated appropriately in relation to specific or general audiences. Such an idea is central to the writings of those prime movers in Language across the Curriculum, Barnes *et al.* (1969) and Martin *et al.* (1976).

The London team would seem to have provided us with a

working formula (varying the audience, vary the function), grounded in theory, for widening opportunities and providing a variety of contexts in school for writing and talking for learning. The very significant role which they ascribe to talking and to writing in draft can be seen to fit with notions of inner speech being the seedbed from which understanding and reasoning develop. Douglas Barnes and Frankie Todd have taken these ideas further and examined learning in groups. Their analysis of talking in groups substantiates claims about the helpfulness and significance of talking-to-learn in learning. A specific piece of research by Cummings (1985) in computer-assisted learning (CAL) in geography supports the value of the talking-to-learn hypothesis and shows how it can lead to a questioning of statements and ideas.

A SMALL EXPERIMENT

Now, a decade or more beyond the early murmurings and formulations (see the Bullock Report, for example) of Language across the Curriculum, other preoccupations shift the emphasis away from such fundamental concerns as how children learn. Language does not appear as a priority in lists of in-service needs for England and Wales. Is the message to be lost, or did it sometimes fall on fertile ground, to germinate and grow? Will the message continue to be spread? I had a recent opportunity to explore the understanding of the message among a group of geography educators.

They were asked to respond to the questions 'What does the phrase "Language across the Curriculum" mean to you?' and 'To achieve application of the "theory" what practical suggestions do you make to your students?' If I take as a benchmark the response which one close to the London group wrote for me, then the responses of the geography group can be analysed for their general congruence in substance and tone to that benchmark. Here, first, is what Alex McLeod wrote:

> Theories of learning are made of fairly slippery material.
> Most school learning is language-saturated. Some people are still tempted to believe that those who have learned the language have taken on the knowledge, and that knowing something requires only that someone has heard something, or read it, and understood it. This theory implies that knowledge, and the language in which that knowledge is expressed, are identical. It also implies that language and thought differ only in that one is capable of

Language and learning

making a noise or marks on paper, and the other is not.

I have come to believe that knowing and understanding have to be active processes, or they are nothing. Something can be learned only when it passes from experience into thinking, and meets and mixes fairly comfortably with what is known already, and is there to be called on, called back into language, when it is needed.

People cannot be expected to learn anything at any time. There are those things we know that we know thoroughly; there are other things which we do not know and which we could not learn, even if a skilled teacher spent a day teaching us. There is, as well, the 'zone of proximal development' described by Vygotsky (1978) and taken up with enthusiasm by Bruner (1986) and others. This is the zone where we are able to learn; in that zone is all that knowledge and understanding and wisdom which is not yet ours, but which we are able to make our own if we engage actively with it.

We may not need the assistance of a teacher, though we probably will. That teacher may or may not be someone who deliberately sets out to teach us; it may be someone who wrote a very good book, or someone who deliberately, or even accidentally, put some new experience in our way. The 'teacher' might well be a group of other learners, who are in a similar, but not identical situation. Not identical, because the zone of proximal development of one person can never be the same for another.

The essence is the active engagement of our conscious thought processes - explaining the new concept or idea or theory to all the old ideas and concepts and theories that we possess already. In school, this active engagement is best achieved by talking to learn, by having the chance to try out in talk an idea which may or not be the one we are looking for, by having one's 'first draft thinking' overheard by another learner. So small groups are usually better than whole classes; student dialogue is incomparably more useful than teacher monologue; the knowledge we have already in our minds must be given the opportunity to meet and mix with the new knowledge. Biologically, the gut assimilates and the eye accommodates. The conscious brain has to do both simultaneously, to take on the new and absorb it, and to be changed itself in the process.

As much as first draft talking we need first draft writing, because both are drawing on first draft thinking. We very often do not realize that we know something until we write it - but we are not Mozart, and we can't expect to produce a faultless finished piece first time round.

Learning has to become systematic, organized and coherent. We may be able to learn about systematic and organized thought from other people, but in the end we can know and mean only what we can construct for ourselves.

(26 January 1987)

The following are replies to the first question which I composed from those responses which seemed in substance and tone close to Alex McLeod's text. The numbers refer to the number of people responding in such a fashion.

Language across the curriculum recognises the significance of language in learning for shaping and developing ideas. (Five people)

There is an understanding that spoken language is as important as, and complementary to, written work, i.e. there is a relationship between 'exploratory' writing and talking and 'finished product'. (Nine people)

Conceptual thinking depends on language development and use - the two are integral. (One person)

Children should be using language to originate, explore, and develop ideas. This can be creative and open up ways of learning. (Two people)

Language is important too in communicating ideas and reporting findings, both orally and in writing. (Two people)

Varying the audience and purpose helps to open up a range of opportunities for writing and talking so that a breadth of existing knowledge and experience can be brought into play. (Five people)

The expressive, creative, aesthetic, and emotional side of development can be explored via language, as well as the more formal transactional kind of thinking and reporting. (Four people)

The 'across the curriculum' aspect is important. Language is one of those areas of development to which all teachers should contribute. (Four people)

One other idea expressed which probably postdates the London group's work and is more closely related to Lunzer and Gardner's work is:

To allow children the opportunities to read and interrogate a variety of geographical texts with a view to facilitating

Language and learning

their understanding of the material contained in the texts. (One person)

It is strongly recognized above that there is a relationship between exploratory talking and writing and a finished product designed to communicate ideas accurately and appropriately; that language is significant in learning; and that varying the audience and purpose provides a range of opportunities for talking and writing so that a greater breadth of existing knowledge and experience can be brought into play. The points additional to the first three most frequently mentioned ideas reinforce and make them more explicit. The summary as a whole gives us a sense that where the Language across the Curriculum message has been picked up by geography educators it is sensitively and accurately understood. However, it was a minority who expressed themselves in the terms set out above, and, for that reason alone, one may feel pessimistic about the likelihood of the message, linked so closely as it is to learning, surviving and spreading.

Those who responded with what is generally taken to be the accepted understanding of Language across the Curriculum were also able to express specific ideas in response to the second question: 'To achieve application of the "theory" what practical suggestions do you make to your students?' There were also congruent practical suggestions made by those who had not necessarily closely identified the Language across the Curriculum message in their first response. The composite reply below lists a range of topics for consideration which match the general message.

> How to stimulate constructive talk in the classroom and field (buzz groups, small groups, pairs, role play, simulations, games, problem-solving activities, decision-making activities, reporting back, listening to the pupils, pupil tape-recording, oral fieldwork, talking to the public, etc.). How to encourage the use of the pupil's own language. (Ten people)
>
> How to encourage pupils to express themselves initially, even if the form of expression is not particularly elegant, accurate, or clear. (One person)
>
> How to extend the range and purpose of writing - expressive and transactional. Setting imaginative tasks changing the nature of the audience, negotiating the tasks, using written work as a starting point, not always as an end point, encouraging aesthetic considerations.

> Consideration of the use of dictated notes, copying, and long periods of listening. Consideration of bias. (Ten people)
> Consideration of the impact of marking and assessment in general upon children's use of language in and out of school and how to respond to work. (Two people)
> Asking teachers about their Language across the Curriculum policy. (One person)

Again, other points derived from Lunzer and Gardner's work on DARTS came through:

> Use Reading for Understanding, DARTS, ideas to help pupils engage with and interrogate a text. (One person)

In summary, I think it may be said that the range and wealth of ideas for putting Language across the Curriculum into action is impressive.

RESPONSES WHERE LANGUAGE ACROSS THE CURRICULUM SEEMED TO BE LESS WELL UNDERSTOOD

The characteristic responses to the first question ('What does the phrase "Language across the Curriculum" mean to you?', etc.) of those who did not seem to interpret the phrase as in the composite positive description above, or who did so vaguely or only partially, included the following:

> The importance of vocabulary and language appropriate to age and ability.
> The accurate use of parts of speech through the expressive use of language in many different contexts.
> Choosing the most appropriate words and concepts for explaining something to young people.
> A consideration of reading, writing, listening, and talking in a geographical context.
> The incorporation of 'language skills' beyond the traditional confines of 'English' where language skills involve written communication, oral communication, and listening skills.

It was not possible to classify vague statements - some even less precise than those above - one way or the other, and indeed it could be argued that the last two above defy

Language and learning

classification. They were taken from responses which omitted any mention of the role of language in learning at all.

For practical suggestions this second group placed emphasis on consideration of level of materials; the Postgraduate Certificate of Education (PGCE) student's own use of language; readability indices; language for slow learners; language for mixed ability; the production of appropriate work sheets; marking for accuracy. None of these matters is trivial or unimportant, and all need to be considered. Such statements, however, are taken as examples from longer responses not grounded in the context of the significance of language in learning or the place of talking in learning.

CONCENTRATING ON WRITING

A second part of this small and very exploratory investigation of the understanding of the Language across the Curriculum message was to have the group comment on the two pieces of writing reproduced below. The style adopted in each case is very different. In the first case the pupil was asked to write a radio programme based on a farming game which had incorporated a simulated radio farming report and, in the second case, to write notes for homework.

Passage 1

> *Homework, 17th November 1986*
> 7.50   'A farmers choice' BBC2
>
>   'Action'
> *Bill Passfield (Presenter).* 'Hello this is another edition of "A farmers choice". With us tonight is Adam Mortimer.'
> *Adam.* 'Hello, I'm a farmer. My farm is an arable farm.'
> *Bill.* 'Just in case you don't know an arable farm is a farm with only crops on it. Why did you choose to make your farm just crops?'
> *Adam.* 'Well Bill, my farm has rich soil which is good for sugar beet, corn and wheat also my land is flat which makes it easier for the machines to plough the land.'
> *Bill.* Does the weather affect the growth of your seeds?
> *Adam.* 'Oh yes, in spring and summer we need both sun and rain. We plant the seeds in December so they grow a little before the frosts in January, February and March. If we don't plant the seeds in December they will not stand up to the cold winds. They will just blow away.'

Bill. 'Thank you Adam ... Here is another farmer his name is John Barron.'
John. 'Hello, I'm the owner of a pastoral farm, a pastoral farm has just animals on it.'
Bill. 'Where is your farm?'
John. 'It is in Worcestershire on the Cotswold hills.'
Bill. 'Why don't you grow crops?'
John. 'My soil is very poor and we also have the steep hill and the tractors have difficulty in climbing it.'
Bill. 'What other activities do you have on the farm?'
John. 'We have hay-making, an orchard, vegetables, looking after the animals and a little crop growing, not for the market but for the animals to eat.'
Bill. 'Um yes. What products do you get from your farm?'
John. 'I get beef, milk, cheese, eggs and bacon.'
Bill. 'Thank you John ... Good night everybody, see you next week.

'and CUT, Good Good one more time'

Passage 2

Upland country farming   20th November

The farm is in North Wales in Gwyned. A man called Mr Roberts owns it. As you get higher up the hills the weather gets worse. It is 3 times higher than sea level. The way things have changed are the men used to walk into market, get up at 2.00 every morning. The soil gets very water logged and the grass doesn't grow properly. He uses the land for sheep grazing. He uses it for that because the soil is poor so he couldn't grow crops. 3/4 of the profit comes from Sheep's lamb.

A significant number of the group saw the first passage, the interview, as an enjoyable and stimulating activity for children, and of a kind which PGCE students should be encouraged to use with pupils. The passage was judged to have a high interest level and to be successful in getting points over clearly. A sense of audience was considered to have influenced the work. Other responses also of significant number commented on the traditional paradigm of geography, even the deterministic flavour in which the interview was written, and the rudimentary concepts of soil and lists of products given. A weakness of the piece, one tutor wrote, was the apparently

Language and learning

insufficient knowledge of farmers' everyday lives. Such knowledge -through a farm visit, for example - might have helped to bring the dialogue more alive. Another tutor felt that it was writing 'dressed up in a progressive format', while another's constructive suggestion was that small groups using tape recorders prior to the written task might have helped to increase the number of concepts and the depth at which they were explored in a written dialogue. There was a tendency in the responses to focus either on the style and 'openness' of the writing or on the limitations of the geography. A very small number did, however, comment positively on the written aspects and less positively on the geography.

The responses to the second piece of writing were short and sparse, for whatever reason, and time may have been one. One very balanced response went thus:

> Equally well written in terms of 'style', etc., but obviously a more formal description.
> Who is it communicating to?
> Does the writer know what is required of her/him?

Other responses also thought there was a lesser sense of purpose and audience than in the first passage and that it was not as interesting to read. Passage 2 was described as conventional and not well structured, even muddled, with possible misconceptions. In a sense neither passage scored admirably from the point of view of geographical knowledge but the first was praised for its liveliness and sense of audience - for its writing style, in other words. It is not surprising that one tutor wrote, 'Are we marking it for Geography or English or both?' Or is it surprising?

However we answer the last question, the results of the invitation to comment on the two passages suggest that research is needed into ways of articulating the differences, if any, which exist between audience-specified and non-audience-specified styles of writing and talking. It would probably not be enough to take examples for analysis but would also be necessary to talk to pupils and teachers for insights into their motivations and meanings.

THE PROCESS OF LEARNING

There was some correlation of response between positive comments on passage 1 and knowledge of the specifics of the Language across the Curriculum message in the first part of the investigation. There could be some grounds for the accusation that some of us (and I am one) who believe in the links the message makes between language and learning are making our responses through rose-tinted spectacles, that our framework determines how we interpret things. This would be equally true of those who use other spectacles or frameworks and make other comments.

The tutor who saw limitations in the conceptual development and range of understanding expressed in passage 1, and suggested that oral dialogue using tape recorders for play-back and interrogation purposes in order to provide opportunities for going over first formulations, had got to the heart of the matter. It seems that he or she was saying that we need to understand learning as a process and to see children's work as stages in that process.

As Vygotsky points to the significance of 'inner speech' so we have to keep clearly in mind the devising of opportunities for engaging in the process of talking to learn or writing to learn as stages on the way. The recognition of talking and writing as exploratory rather than as necessarily finished product is important.

Perhaps a logical step on from deliberately created exploratory talking and writing episodes is to have groups of children interrogate their own texts such as the passage 1 quoted here. I remember suggesting to the student who set the homework which led to passage 1 that it, and examples like it, could be acted out in class and made the subject of class discussion. This would have been one way of going back over the text, a way of looking at it again and providing a chance for reconceptualizations. By such a strategy children would be 'assessing' their work at that exploratory stage. We move in this way beyond the simple formula of 'vary the audience, vary the function' to a closer encouragement of the learning process as articulated by the London team and those who informed them.

> The relation of thought to word is not a thing but a process, a continual movement back and forth from thought to word and from word to thought. In that process the relation of thought to word undergoes changes which themselves may be regarded as development in the functional sense. Thought is not merely expressed in words; it comes into existence through them. Every thought tends to connect something with something else, to establish a relationship between things. Every thought moves, grows and develops, fulfils a function, solves a

problem. This flow of thought occurs as an inner movement through a series of planes. (Vygotsky, 1962:125)

Some of the reactions to the pupil passages reminded me of comments sometimes made by PGCE students when I begin each year to broach the idea of language and learning. From time to time, when I suggest writing tasks with a specific audience, for homework perhaps, and try to ground these in ideas about the learning process, puzzled, uncomfortable looks appear on faces. I am asked, 'But is this geography?' 'Will the kids see this as geography?' We then move into discussion about teaching facilitating learning, about learning through geography, and our conceptions of what it is to teach. To those who remain least convinced the only way through is the challenge of 'Give it a try, see what happens to the writing' - and I mention writing here, not talking, for no other reason than because of the constraints placed on student teachers and their tutors.

PGCE students usually find the results of such 'experiments' rewarding. Their assessments would be assisted by the findings of some in-depth study into the differences between process-nurtured writing and the more conventional notebook record. They find Robert Hull's (1985) work helpful. In many ways his book *The Language Gap* is a 1980s reformulation and rediscovery of the Language across the Curriculum ideas. He focuses particularly on the conceptual and experiential gap between teacher and pupil. Some of his points are similar to Milburn's (1972).

THE PUPIL, THE READER, AND THE TEXTBOOK: ANOTHER COMMUNICATION GAP

The interface between the pupil, the reader, and the textbook is another dimension of language in schools which is receiving attention. It focuses not on the language between pupil and teacher but on the dialogue between pupil and text. Who is the text addressing? This is one of the key questions in this area of research relevant to language and learning and it links with the ideas of those people who emphasize the concept of interrogating the text.

Carol Robson's (1983) work on textbooks, reading, and reading development includes a case study in geography and can, generally speaking, provide geography teachers and educators with an introduction to the area which is useful and helpful. The focus is one about which, I would guess, we are fairly ignorant. If Language across the Curriculum ideas are in early to mid stages of diffusion, then concepts such as the implied reader (i.e. what kind of a reading is implied by the text) and the relationship of the reader to the text have scarcely begun to be talked about among geography teachers.

MODELS OF READING AND COMPREHENDING

Robson grounds her work in models of reading and comprehending which she reviews for strengths, weaknesses, and appropriateness to her purpose. A skills view of reading and comprehension is generally found wanting, since reading to understand and answer questions is a test-dominated view of comprehension. All meaning is held to be in the text, elements of which 'produce a correct meaning for the good reader armed with questions', whereas 'the flexible reading behaviour of experienced readers is a complex activity employed in pursuit of their own learning purposes' (Robson, 1985:15).

The data-processing model of comprehension is also found wanting, though to a lesser extent. Developed by cognitive psychologists, it rests on assumptions of a close relation between the structures of the text and the knowledge and memory structure of the reader. Comprehension is viewed in this model as an interactive process. There is a helpful focus on the knowledge and the expectations of readers as vital contributors to their understanding. The model, however, overlooks how the text might influence the mind of the reader, and Robson does not find the model helpful for the development of comprehension at the reader-text interface. She suggests it does not allow for a range of possible reader responses.

A psycholinguistic view of effective reading developed within the context of an holistic approach to learning is the third and favoured model to be reviewed. Based on the work of Smith (1982), the Goodmans, Kelly and Bruner, such a view envisages a reader coming to the text with intentions and expectations. Both Smith and the Goodmans describe readers as actively interpreting the text, bringing to it prior knowledge which helps them to predict the surface structures on the page. Reading is an exchange of meanings with an author. Smith assumes, for example, that comprehension takes place within the context of 'shared conventions', a writer choosing a register appropriate to an audience's expectations. This group of people may be said to have 'transactional' views of reading.

ROBSON'S INVESTIGATION

It is within this third model that Robson considers she has a framework for beginning to investigate and understand how readers cope with reading and where they focus in on a text. She formulates three main questions:

1. How appropriate are their expectations to the text they are reading?
2. How effectively does the text extend them, so that learning occurs?
3. How do their expectations and understandings of the text compare with those of their teachers?

Narrow concepts of readability to do with sentence length and word counts are not found adequate, though features of discourse which evidence *organizational* characteristics (frequent headings, placing of central ideas, advance organisers, etc.) and characteristics of *cohesion* (backward and forward reference, use of conjunctions, etc.) are valued.

Readability and comprehension need to be seen as a function of the reader-text interaction, according to Robson, and not of the text alone. Narrow views of readability lead to the need for a view of the reader-text relationship which will be more helpful and holistic as an approach to the research questions. Culler's theory of 'literacy competence', defined as an ability on the part of the reader to identify various levels of coherence and set them in relation to one another, and the implied reader concept which *focuses attention on the way texts direct themselves to the reader*, are judged to be helpful analytical concepts for finding answers to text-reader, text-teacher expectations and relationships.

In her investigation (1) Robson interviewed staff to ascertain what meaning they saw and expected pupils to find in texts; (2) the texts were examined for their implied reader or lack of it; (3) lower sixth-formers were interviewed to find out how they found the reading.

The following direct quotations from Robson's thesis give the flavour of her undertaking in the three areas and point towards the significance of teachers more sensitively and fully coming to appreciate likely pupil-text relationships. The first is an account of an interview with a teacher and Robson's interpretation of this event. The second contains the interview with D., an average student, and should be read prior to and in conjunction with the third, Robson's interpretation of the interview.

1 The teacher, the text, and the essay: Robson's interpretation

The teacher's comments, Robson writes, 'show predictably that her expectations have meshed readily with the implied reader constructs of the text' ('Forest and forest products' in E.H. Cooper, *An Introduction to Economic Geography*, London: University Tutorial Press, 1986). The teacher, Mrs G., states:

> It's about the Forest Industry and it's developed as a topic in clear sections. It describes the main timber resources of the different forest belts, their problems, and exploitation, and conservation, and the kinds of industrial development that result and their problems - many of which, like the competition from synthetic products, you get from using tables. The different forest zones relate to developed and developing countries and we've emphasized this division in dealing with Population and other topics recently. We've also done a lot of analysing maps and interpreting statistics in the Population topic which will help them here.

'The opening sentence,' Robson continues:

> indicates that Mrs G 'recognizes' a *systematic* textual structure and, together with the second, shows clearly that she *integrates* all the sections and components of the text, synthesizing them as a topic at an abstract level. Her own constructs lead to particular emphases and re-classifications of the chapter into 'forest industry', 'problems' and 'developed and developing countries' - subject-specific concepts in her usage. The last two sentences demonstrate her assumption that previous teaching will have provided the students with appropriate knowledge and skills which can be brought to bear on the text. Not surprisingly, the essay title, given 'to ensure that the chapter has been read and carefully understood', reflects her whole approach, taking for granted that the pupils' reading will echo her own. It asks them to 'write an essay on the Forest Industry of the world, with particular reference to the distribution of both forests and industries and their characteristics and problems in (i) developed and (ii) developing countries.' (p.81)

The extent to which pupils complied with her expectations can be seen in the second and third quotations:

Language and learning

2 Interview with D., an average student

<u>The approach to, and focus of, reading</u>

T. Now ... how did you get on with reading this?
D. *I read bits of it, then I wrote about them, then I read bits again and wrote about them. I didn't read it all through in one go.*
T. Did you want to read it?
D. Well ... I read it because I had the essay to write, but I did think it was quite interesting - more so than some of the work, so I think I would have read it anyway once I started ... I didn't find it boring, but I found I was losing track of it a bit as I went on.
T. Did you know anything about it as a topic?
D. I knew vaguely.
T. What do you think you got out of it? What was it about?
D. Well, I think the main thing that struck me was how the forests had been used up so quickly ... and how most of the forests in the world had been used up ...
T. Right ... Anything else?
D. The time ... The amount of time it took for a large tree to become mature so you can use it ... so you don't actually live to see the result of your planning and so it's that much more difficult to plan a forest and work out what's going to be used when. I hadn't really thought of that.
T. Anything else that seemed to you to be important?
D. Well, for my essay I chose the different types of forest and whereabouts they were situated and then there are the different types of uses of wood. Pulp and paper. I knew that was important. We hadn't actually learned about it, but I'd seen a television programme, but I was surprised the way it was made. I knew it was sort of mulched up, but not about the expense of it and that it was done chemically as well. They put in chemicals to get rid of the varnishes. It's much more expensive, but makes a much higher-grade paper ... And rubber ... I knew about rubber and latex and that - 'cause we did it for O level and in second year, but I couldn't have pinpointed specific places.
T. Right. Did anything else strike you?
D. Yes ... Looking at the figures, some of them seem incredibly ... large. You wouldn't expect nearly half as much to be used, and the amount in one year! ....

You wouldn't think so much timber could be produced in one year.
T. So - the sheer amount of it was very striking. Anything else?
D. Well, I expected to come across the forests in Siberia and Russia and that area, 'cause I've seen films of it, but it says in parts of Africa there's a lot that hasn't been used and it's not being developed and I found that surprising.

Awareness of sections and headings

T. Now, turning again to how you said you read this, you said you read 'a bit' and then wrote your essay. How far did a bit go?
D. Usually its, sort of, a heading. Like in 'The Location of Pulp and Paper Mills' it was about a page ... just different headings ... different blocks ... Normally, with Geography, there's always headings inside the chapters and you know what's going to happen. There's different kinds of heading ...
T. What different kinds of heading are you aware of?
D. *Well, there's the main heading, like ... er ... 'Distribution' - where they are -* and then it goes onto 'Conservation' and *then 'Industry' - the kinds there are.* They're the different main blocks, and then there's little side headings, like 'Tropical Hardwood Forest' and so on, *and they're all separate bits.*
T. When you're reading, do the different kinds of heading help at all?
D. Er ... *When I'm reading, I don't so much think of them as different kinds, I think of them as following on, one type of tree and where it is and then another - and I picture them,* like with *'Tropical Hardwood Forest'* I'd be thinking of low-lying ground, with monsoon-type climate. I suppose I see it as jungle more than anything else. It's usually very wet and very hot ... *It's not so much a place - an* area - actually very dense [laughing]. *It's what it looks like in your mind,* with trees everywhere.
T. Did the 'Hardwood' form part of your picture?
D. No - I didn't think about that really, till later on they mentioned furniture wood, like mahogany and teak. It tells you - page 194, the second paragraph. Then I just thought of furniture [laughing].

Language and learning

T. Were any of the other headings significant to you?
D. Well, the *'Temperate Hardwood'*. I thought of more normal trees - ordinary - like oak, like you see in English woods and in places like Europe where it's not so hot and where the forests are used more. *Then 'Softwood Forests of the Northern Hemisphere'*. I thought of Russia, 'cause I knew they had softwood, so I immediately thought of Russia before I read the section. I didn't have a very good picture ... but I've seen a few films of Siberia and I think of the pines ... that's a main image.
T. Do you think of 'softwood' at all?
D. Not specially ... though I suppose it's used in more delicate processes, - rather than just ordinary planks ... I think of it more for furniture again. I know a lot of that's hardwood - but for more decorative, for more sort of intricate furniture.
T. Was there any heading that was particularly unhelpful?
D. Yes! 'Softwood Forests of the Southern Hemisphere' [reading it]. That just sounds long-winded.
T. Now, thinking of this first section, still, for the moment, are there any other kinds of headings that you notice in your reading?
D. Well, there are smaller ones - different areas - Scandinavia, and then it goes on to Canada and the USA.
T. Was giving the countries and areas helpful?
D. *No, I don't think of the countries much ... With Scandinavia and Norway I do, because of the Christmas tree in Trafalgar Square*, but I don't really think of the places.
T. So the countries don't seem important?
D. They are important, but I don't take much notice of them, 'cause they're not what I'm trying to get hold of. *I'm trying to get hold of what they look like and where they are generally and how many there are*, rather than the actual countries - I might p'r'aps remember as an example that there's some kind of forest in a particular country, but not much ... *You can't take in all the sections, because they get to be too many to make sense of, so you decide what's most helpful for you.*

Reading of 'location factors in ... paper industry' section, p.206

T. Were there any headings or sections anywhere else that were particularly helpful for you?
D. Some in 'Industry' were helpful, like ... 'Location Factors in the Pulp and Paper Industry'. They don't usually say something like that that helps you see what it's really about. They'd more likely say 'Pulp and Paper Industry' and you'd have to work out that it was about location factors, and what they were, for yourself. There are four factors involved.
T. What does 'location factors' mean to you?
D. You think of why a certain industry is situated where it is - the natural resources and relief of the area, that kind of thing, and the amount of space that would be needed ... The things that affect why it's there. These were mostly things I knew. Like, I knew about access to forests [laughing]. You couldn't have one in the middle of a town, you know! And I realized it would be a heavy consumer or electricity or whatever, and I knew that water was used and nearness to water was important but I didn't realize it used so much - 100,000 gallons for only a ton of paper - and I thought that it would be a large employer of labour, but it isn't. I suppose it's because it's mostly mechanical.
T. What did the 'Industry' section mean to you - how did you think of it?
D. Well, *I was noticing what the industries were and then mainly how they do it and how much there is.*
T. Did the industries link to the kind of wood and distribution in the first section at all?
D. No - not really. I suppose, if you knew enough about it, you could link it back, but I tended to think of it as quite separate.
T. You mentioned rubber before. Did that link to any kind of forest for you?
D. No. I wasn't thinking of that ... I thought more about how it was done. I know the wood's cut at an angle and ... er ... that's rubber tapping and there's a little basin thing that collects it. I didn't relate it to places. I was thinking of another TV programme mainly, with rows of trees, cut at an angle of 45°, and the little drip trays that they collect it in ... I think maybe it was in India or Ceylon or somewhere like that.

Language and learning

Reading of 'methods ... of forest management' section, p.204

T. Now we've been talking about headings or sections that you'd particularly noticed or felt were helpful. Is there anything else that you were aware of as you read and that you'd like to tell me about before we move on?
D. Well - er ... there was the bit about Conservation [looking], p. 204. There were little headings that picked out points. 'Forest Establishment', I was thinking of the renewing of the trees that had been used and making up the stock and er ... then 'Selective Cutting' - cutting only parts of the forest at a time so you always save some. When I first read them I took them as all separate, but then after I'd read them they seemed to connect up to conservation and they were numbered too. I already knew about some of them and about the idea of conservation.
T. How did they connect up?
D. I realized that obviously 'Conservation' was the main heading - protecting the forest - and this was the different ways that people were trying to conserve the forests, so it linked to that heading.

Awareness of maps and tables

T. Right, now, did the maps and tables figure at all in your reading?
D. Well - I usually read a block, write about it, and then look at the maps and tables.
T. Were there any that you particularly noticed when you did come to look at them?
D. The one on p. 195 was more confusing than anything. There were so many different shadings and there was writing everywhere. I was too confused to take much notice of it, really.
T. Did it clarify anything at all?
D. *Well, it gave you some information, but a lot of it was too complicated and the useful bits you'd been told already.*
T. What about the tables, did any of those strike you as helpful?
D. Well, I thought the little one on p.199 was interesting ... The numbers! I couldn't get over the size of them! They seemed unbelievable, the amount in one year. I'd have thought it would have been about ten years.

T. Any other tables?
D. I thought table 46 on p.201 was useful, 'cause you could compare and you could see clearly which were *the most important* and the *least important countries* for timber.
T. Did you think of any reasons for countries being where they were on the list?
D. No, I didn't, actually.
T. How do you think of a table? Does it tell you things that aren't in the text, or things that are the same, or what?
D. *I think of it as a précis of the text more than anything else.* You don't get the actual figures in the text, but they put it another way. The table gives you an outline – I usually notice the top three or so countries or whatever that have most, and p'r'aps the bottom one, *so you get a general idea, but those things are in the text as well.* Some tables are very complicated, and they're just more detail than you need, so I don't bother with them.
T. Which ones didn't you bother with?
D. Well, the 'Imports' one - and the 'Production of Wood Pulp'. And I didn't think the 'Rubber' was very relevant either - with the smallholdings. It seemed haphazard, the way it was laid out and complicated.
T. And you looked at tables at the end of a section?
D. Yes. It means more. If it says 'Fig so-and-so' in brackets, I don't look then, I wait. I might look back to the point where it told you, when I look at the table, but usually I don't. *It doesn't really relate up all that well and it doesn't always tell you to look, anyway. I think you look if you want to.*

Final impressions of topic and text: criticisms

T. Right, now when you came to the end of it all, what seemed to be clear in your mind? How did it make sense as a topic? Did it make sense?
D. Well, I think I understood the blocks of it quite well ... The distribution is obviously important and the different types of conserving was quite good, 'cause there's not just one way and, er, with the industries ... there's what they are and how you do them.
T. Were you linking the industries back to types of forest, types of wood, at all?

Language and learning

D. No, I wasn't thinking of whether it was coniferous or so on that was there forest-wise. I was visualizing the different industries and how they were organized.
T. And do you feel you've got hold of the topic clearly?
D. Much more so than when I started, but it could be clearer. I'm a bit confused about some of it, really.
T. Why is that, do you think?
D. Well, I was surprised at all the different types of wood, but I thought that all the headings - 'specially in the first section - spread it out too much ... *I thought it should have been in a closer block, 'cause every time you turned a page you saw another heading*, and *it seemed an awful lot of different bits.*
T. You mean it shouldn't have been subdivided, or what?
D. I was glad it was when I was reading it, at the beginning anyway, 'cause I thought it was going to make it clearer ... but looking back it went on - and on - and on ... *I think they went into the woods too complicatedly. They could have just said 'Softwoods' and 'Hardwoods' or whatever, and then told you what they were and where they grew* ... I found it rather confusing ... *They didn't seem to stick to one kind of heading ... They seemed to change their minds* and I'm not really sure how the different sorts go together.
T. So you found it hard to relate the different headings to each other, to ... er ... to form a section, is that it?
D. Yes, I did ... They fitted together in bits, but not all of it.

Awareness of 'problems' and 'developed/developing' countries

T. Did, er, did the essay help you to relate the different sections at all?
D. No - not really ... 'cause that was about the Distribution - that was the first part and then the Industries, that was part of it too. Problems ... that was difficult. I think that was really the problem of conservation.
T. Were there any other problems that struck you?
D. Well, I looked through, but I thought that was a difficult part to answer. It didn't tell you about many problems. I think it was the demolition of forests more than anything else.
T. What about Developed and Developing Countries?
D. Well, I know some of the forests are ... like temperate

ones are in the developed and like the tropical ones are in developing, but I didn't really think about it much in that way ... 'cause there were a lot of bits already to keep track of, like I said.
T. Right, that's fine. I think that will do, D., thank you.

3. The pupils' experience of reading

Unlike the teacher, Robson finds that:

> none of the girls had the competence of the implied reader, and in particular none brought to the chapter a sufficiently developed awareness of 'classifications' or the ability to interpret maps/tables and relate them to text to enable them to integrate the different sections and levels of the topic as the teacher expected, and each was conscious, at some stage of her reading, of being frustrated by the text. However, each focused on some of its contact features in ways appropriate to her own expectations and learning purposes, thereby developing localized understanding and formulating or confirming impressions about Geography texts.

'D., an average student'

> Robson interprets D.'s responses in the interview thus: D.'s technique of reading a section at a time and then writing a portion of her essay suggests that the organizational structure of the text does not activate expectations of a hierarchically classified topic, with interrelated sections, but a *sequence* of separate 'bits' or 'blocks' loosely grouped at a fairly concrete level by the three main headings which organize them descriptively rather than as a true classification. As she moves from section to section, unable in most instances to supply the omitted bases of the classification, she dismisses some 'bits' as incomprehensible. Those on which she focuses make sense to her to the extent to which they confirm, elaborate or challenge a mental *'picture'* - a generalized description, or narrative fragment - often originating from a film, and it is at this level that she tries to extend her learning. She says - with reference to the forest types in 'Distribution' -
> 'I'm trying to get hold of what they look like, where they are generally, and how many there are ...' In the

Language and learning

'Industries' section she is concerned with 'how they do it and how much there is' in each case, and focuses on Pulp and Paper-making and Rubber, where she already has clear visual images. Since she is grasping the above 'bits' so concretely, the text's alternative method of classifying by countries cannot be incorporated into her understanding except in rare instances (like Norway and the Christmas tree in Trafalgar Square) where they relate to a visual image. She therefore adopts a deliberate strategy of elimination and selection as she proceeds: 'You can't take in all the sections because they get too many for you, so you decide what's most helpful for you.'

The analysis by Robson proceeds:

What is 'more helpful' continues to be basic facts, so that D. continually fails to establish different 'levels of coherence'. Her failure to integrate maps and tables is significant here. They are considered irrelevant to the extent of being referred to only after the text has been read and the essay written, and she expects them merely to repeat, or 'précis', what is in the text. Her level of competence only enables her to see tables as indicators of 'least important' and 'most important' items, and maps and tables only as repositories of basic facts. (It is the sheer *amount* of timber, etc., produced which constantly amazes her.) Therefore, given no guidance to develop further competence by the text's contact features, which, at best, seem to offer only an *optional* directive to look at maps/tables, she follows her own learning needs and either simplifies them to a few facts which usually duplicate the text, or rejects them totally as too complex. Sometimes she does both! In any case, with the text's assistance, she reinforces her view of their irrelevance.

D. only successfully relates different levels of coherence in two brief instances ('Location Factors in the Pulp and Paper Industries', p.206, and the methods of managing forests, p.204) when the organizational features of the text exactly mesh with her expectations, enabling her to use headings to anticipate the way facts will relate and to organize these successfully as a category which 'explains' the heading. In both cases, though numbers are a helpful contact feature, D. is confirming or only marginally extending knowledge already organized at this level of abstraction.

D.'s final understanding with which she is fairly

satisfied is still largely at the basic level already described, with classifications of the three main sections unrelated and no reclassification in the teacher's terms. However, her moments of accord with the implied reader, and an increased sense of there being 'an awful lot of different bits', also make her dissatisfied with the degree of coherence she has established. She begins to identify features which would make the text more 'readable', thereby indicating both the beginnings of discourse competence and the textual characteristics which would enable her to improve it. She thinks that headings should be more explanatory and ways of classifying simpler and more consistent, leading to fewer sections. (pp.82-4).

The extracts fill out a sense of what an implied reader means and levels of coherence which are achieved by one reader in her interaction with a text as she tried to make meaning. Robson felt that the teacher's ignorance of the pupil's reading struggles made her, the teacher, critical of the pupil's failure to demonstrate a competence which she did not have and which the text did not help her achieve. The pupil herself gave evidence of *negative* learning about geographical discourse; she avoided using maps judged by her to be irrelevant and used tenuously related sections.

In conclusion Robson makes the following points:

1. The investigation shows that a reader does not encounter a text that 'contains' meaning; she supplies subject, cultural, and textual knowledge as she reads. The experience of making sense of a text is modified or reinforced by the textual encounter and develops consistently.
2. A-level texts often confront pupils with unfamiliar discourse. They are very actively involved in the lower sixth in developing discourse competence. They consciously identify and distinguish discourse components and how they relate. Teachers are concerned with meaning; pupils with how to construct it. The development of competence at this stage involves a transition from narrative-dominated strategies to strategies for relating the more abstract levels of coherence found in the subject discourses.
3. Understanding of what is read is likely to be uneven. Some of the learning is negative, given the inappropriate texts.
4. The readability of a text would seem to be closely related to its capacity to encourage the development of discourse

competence. To achieve this it needs to employ 'contact' features that are personal and include 'redundant' explanation and 'metalingual' guidance and to guide readers to identify and relate different components and levels of discourse, enabling them to progress from more concrete levels and familiar strategies (e.g. visualizing, following chronological sequence) towards more abstract ones (e.g. synthesizing 'classifications' and tracing argument).

As means towards such ends, Robson considers, if pupils

> compared their 'acts of construction' with other readers, and all [had] re-examined and discussed how they arrived at their own versions, in an atmosphere which encouraged them to see that the 'act of reading' was as important as the text ... they could possibly [have] come to realize the power of their expectations, [and their sometimes distorting effect], and [have] learned to distinguish further contact features and their mode of functioning.

WHERE WE ARE

The work which has been done to try to understand better the significance and relationship of language to learning in the process of learning is, I think, impressive. Teachers of geography have a number of theoretical groundings to which they can turn for suggestions or explanations of puzzles and difficulties encountered in the learning process in the classroom; they have bodies of research evidence to which they can refer and add; and they have suggestions or strategies which may help to overcome difficulties in the interface between language and learning. In this chapter, I have given some account of areas which seem to me of helpful relevance to geography teachers and those about to become geography teachers. It seems that we have to develop a greater confidence in the 'talking and writing to learn' hypothesis and increased competence in managing to put congruent strategies into practice while turning our attention to other aspects of the 'language gap', particularly that between reader and text. I hope this chapter will go some way towards encouraging and stimulating thought and work in language and learning.

REFERENCES

Barnes, D., Britten, J., and Rosen, H. (1969) *Language, the Learner and the School*, Harmondsworth: Penguin.
Barnes, D., and Todd, F. (1977) *Communication and Learning in Small Groups*, London: Routledge & Kegan Paul.
Bruner, J.S. (1986) *Actual Minds, Possible Worlds*, Cambridge, Mass., and London: Harvard University Press.
Cummings, R. (1985) 'Small group discussions and the microcomputer', *Journal of Computer Assisted Learning* 1:149-58.
Department of Education and Science (1979) *Aspects of Secondary Education*, London: HMSO.
Hull, R. (1985) *The Language Gap*, London: Methuen.
Lunzer, E., and Gardner, K. (1984) *Learning from the Written Word*, Edinburgh: Oliver & Boyd.
Martin, N., D'Arcy, P., Newton, B., and Parker, R. (1976) *Writing and Learning across the Curriculum*, London: Ward Lock.
Milburn, D. (1972) 'Children's vocabulary', in N. Graves (ed.), *New Movements in the Study and Teaching of Geography*, London: Temple Smith.
Piaget, J. (1976) *The Language and Thought of the Child*, London: Routledge & Kegan Paul.
Robson, C. (1983) 'Making sense of discourse', unpublished M.A. thesis, University of London, Institute of Education.
Slater, F. (1982) *Learning through Geography*, London: Heinemann.
Smith, F. (1982) *Writing and the Writer*, London: Heinemann.
Vygotsky, L.S. (1962) *Thought and Language*, Cambridge, Mass.: MIT Press.
 (1978) *Mind in Society*, Cambridge, Mass., and London: Harvard University Press.

Chapter Two

WRITING IN A HUMANITIES CLASSROOM

Daniel Lewis

INTRODUCTION

This case study focuses on the writing of two bilingual pupils, Fatima and Ferdinand, as they study humanities in the third year of a secondary school. The two students differ significantly in their success at school, including their writing. Fatima is seen as a model pupil, with enthusiasm, intelligence, and work of high quality. In contrast, Ferdinand is a low achiever, even though he is conscientious and tries to keep up. Three factors that influence their level of development are discussed in this study. First, there is the effect of the classroom environment on the writing process. Does it encourage or hinder writing development? Second, there is the ability of the teacher to recognize the individual writing needs of pupils such as Fatima and Ferdinand. Examples of their writing are analysed and there is a discussion about the stage of development they have achieved. Third, I examine the influence of personal and social factors on their writing, with special emphasis on their personalities and language background. Within a social context, the personal histories of Ferdinand and Fatima will lead to an individual response to the writing demands of the school.

THE WRITING ENVIRONMENT: INSIDE THE CLASSROOM

Imagine you are in a third-year class which is studying humanities on a Friday morning. The teacher spends the first twenty minutes talking and asking questions on the topic 'Rich world, poor world', using a booklet which has information and exercises. Fatima and Ferdinand are listening carefully, both contributing to the teacher-led discussion. Fatima asks questions but Ferdinand is more hesitant, waiting for the teacher to involve him in the lesson. The teacher tells them to begin working and they both start immediately. The first set of questions involves copying sentences and filling in the missing words:

Work sheet from a humanities lesson

1. Copy out the following sentences and fill in the missing words.
(a) The real division in the world is between __ and __.
(b) When countries produce more and more goods and services for their inhabitants we say that __ __ has taken place.
(c) The continued growth in production and rise in living standards has passed by __ __ of the world.
(d) The top __ of the world's population earns __ of world income.
(e) The most important problem for poor countries today is to eliminate __ __.
(f) A large part of the now poor world used to be __.
(g) At first __ were seen as sources of exotic goods, e.g. spices from __.
(h) The imperialist powers killed off the __ __ in the colonies.
(i) There was more profit to be made from selling __ __ than from selling __ __.
(j) One in __ of the world's children go hungry.
(k) For every well fed man, woman, and child in this country there are __ hungry ones in the poor world.
2. Explain the meaning of the following words or phrases: Economic growth, Standard of living, Absolute poverty, Profit, Industry revolution, Inequality, Colony.
3. When did the gap between rich and poor countries start to develop?

Most of the sentences in the questions are similar or the same as the ones in the text. Fatima finishes them after twenty minutes but Ferdinand only gets up to question 1(f) by the end of the one-hour lesson.

Ferdinand is sitting between two other bilingual students, Tahir and Zahid. He knows how to do the questions and searches the text for the answers, but the other two are unsure, so they ask him what to do. He resents the intrusion.

Ferdinand. Go away.
Tahir. What do I do?
Ferdinand. Easy, you find out what it ... I thought you ... I thought sir explained it. Go away.
Tahir. Do we write out the question?
Ferdinand. You're supposed to fill in the gaps, what do you think!
Tahir. I don't want to fill in the gaps.
Ferdinand. Well, unlucky, you have to.

At one point Ferdinand covers his work to stop the other two copying his answers. Occasionally they refer to the work in their conversation but not very often. There is a long discussion about the price of sweets when Tahir says he is hungry. Interaction with the teacher occurs when he notices that Tahir has not started writing and, later, when they are unsure about the answer to question 1(f), where the wording is not exactly the same as the text.

The low interaction with the teacher and the reluctance to share with other pupils contrast strongly with Fatima. She has the ability to switch from focused concentration on her work to high levels of interaction. If she is unsure about something she does not hesitate to ask the teacher for assistance. As she is usually ahead of everyone else, many pupils go to her for help, which she is usually willing to give. In this lesson she answers question 2, which involves more than just extracting sentences from the text. She looks up work done earlier in the year, remembers what the teacher said at the start of the lesson or, if necessary, asks the teacher questions.

Ferdinand starts question 2 the next lesson, missing out the rest of question 1 and failing to do some homework writing about a graph. He finds the question difficult and only says what two of the terms mean:

Economic Growth means when countries produce more and more goods and services for their inhabitants.
Profit means when you buy something for £1 and sell it for £2 you will have a profit of £1.

The definition of 'economic growth' comes from the booklet and of 'profit' from the talk by the teacher. In his answer to question 3 he attempts to explain why the gap between the rich and the poor countries started to develop:

it started when the countries took the good from their colony's and sell them again

He is not very clear and the details are implicit. Perhaps he assumes that the teacher knows what he is trying to say, which, of course, is true. Looking at the original, many features make it seem a poor piece of writing: untidy handwriting; failure to spell 'goods' and 'colonies' correctly; short length; and using verbs in different tenses.

ENCOURAGING WRITING DEVELOPMENT

This brief description introduces us to Ferdinand and Fatima and some of the writing activities they have been asked to do. Is their writing development encouraged? The Writing across the Curriculum Project (Martin *et al.* 1976) and the National Writing Project (Richmond, 1986) provide a framework for evaluating these lessons, and a selection of key ideas would include:

1. Encourage a range of writing:
(a) Different kinds, e.g. personal, descriptive, creative, discursive, analytical.
(b) Different forms, e.g. poetry, plays, newspaper articles, letters, stories.
(c) Different audiences, e.g. teacher, pupils, friends, parents, newspapers, firms.
2. Encourage an open writing:
(a) Pupils able to write in their own words at length.
(b) Opportunities for drafting and rewriting.
(c) Opportunities for collaborative as well as individual writing.
(d) Selective and sensitive marking by the teacher, encouraging pupils to become their own assessors.

One general idea is that a range of writing should be encouraged in terms of kind, form, and audience. Pupils learn to write in different ways for different purposes, and the diversity helps accommodate their range of interests and needs. Another idea is that an open process of writing should be encouraged. This can be achieved by allowing children to write in their own words, sometimes at length, with the opportunity to see their first attempts as a draft that can be revised after discussion with the teacher. Pupils develop a critical sense of their own work and others', in an atmosphere of discussion and mutual support. These two key ideas will be related to my observations of the work done in humanities.

Writing in a humanities classroom

THE RANGE OF WRITING

The lessons I observed were heavily restricted in the kinds and forms of writing and the audience written for. In the first exercises the questions use the same phrases as the text, so all the pupils have to do is a word search to fill in the blank spaces. Very little thinking is needed for this activity, and the writing involves no originality. In the second exercise more thinking and reading is needed because the answers are not so easily extracted from the information booklet, but the writing is still very limited. To find out whether these kinds of writing were typical, I collected examples of the pupils' work over the previous humanities and five other subjects, as shown in table 2.1. The categories 'Simple writing' and 'Comprehension' apply to activities such as questions 1 and 2 of the work sheet, as already described. The other categories are different ways of allowing pupils more opportunity to write in their own words at greater length.

The table shows that in English lessons there are various types of writing, especially descriptive, but in humanities, and especially in science, there is a shift towards simpler writing tasks and copying, as well as comprehension and summaries. Humanities and science have no personal or creative writing. There are difficulties with the categories used, which overlap, and the adding up of pieces of writing, which vary in length, but the general trend is similar to that found in the survey by Britton *et al.* (1975), with humanities and science dominated by restricted forms of transactional writing, which is to do with information and ideas. Their findings on audience were also confirmed by my survey; the dominant audience is the teacher-as-examiner and, to a lesser extent, the teacher-in-dialogue-with-the-pupil, the latter mainly in English. There was no evidence of writing for the other pupils in the class or for people outside the classroom.

WRITING AS AN OPEN PROCESS

There were some opportunities in humanities for the pupils to write longer pieces in their own words, and an example is analysed in the next section. The majority of writing, however, involved short answers to a list of questions where there was little opportunity for developing writing skills. Drafting and rewriting were normal practice for English lessons but did not occur elsewhere. It was interesting to see Ferdinand's English folder, which shows that he takes care and

Table 2.1 Types of writing in different subjects

Subject	Copying	Simple writing	Comprehension in own words	Summary description	Personal opinion	Creative	Other
English	–	3	2	8	5	3	On a map, exercises, similes, limericks
Humanities	2	11	8	7	–	–	Maps and diagrams
Biology	8	5	4	2	–	–	Diagrams
Physics	5	8	2	2	–	–	Diagrams
Music	4	–	–	2	–	–	Drawings
Design	4	–	–	2	–	–	Notes on designs

pride in his work, qualities not in evidence in his other subjects.

Observing the way children worked together in humanities, I noticed a range of activities by the pupils, including working as individuals, sharing ideas, and copying each other. If the writing tasks mean that all the children can legitimately write the same sentence for an answer, or select the same words, and the motivation is to get the right answer, then sharing easily shifts into copying. Collaboration was a casual affair, occurring by chance as children worked quietly on their work sheets. There was little encouragement of any kind of group activity. By comparison, in English the tasks were more open and there was more purposeful interaction and group work.

In humanities the teacher is usually the assessor. For Ferdinand the evaluation of his writing is a sensitive issue; he is hesitant about showing his work and likes to keep it at home. This suggests that he has been overcriticized in the past. In English, however, he can discuss his writing with his teacher and feels more secure. In humanities the teachers do not overcriticize, as they are mainly concerned with the meaning rather than its mode of expression. This avoids the negative and inhibiting marking that can often occur but also means that no positive help is given to improve writing. For Ferdinand, then, the process is fraught with difficulties. For Fatima the writing is easily done but there is a lack of opportunity to develop her writing further.

TEACHER ATTITUDES

Why is the range of writing so limited and the writing process so restricted? For a number of teachers the ideas for writing development listed on p.42 are too idealistic. They say that short comprehension exercises ensure that the children read a text or source material and have to tackle a full range of questions. Copying or extracting words and phrases to answer questions means the children have an accurate record for memorization and testing and the phrases copied provide a suitable model for answering examination questions. The simple questions on work sheets give the many low achievers in the school a chance to be successful. Extended writing can complement but not replace the shorter exercises and raises some problems. If there is drafting and rewriting it means that less of the syllabus is covered. Also, longer, open pieces require more marking because children miss out information or get things wrong. For some, the demand to develop writing skills in humanities is seen as the consequence of poor English teaching. For others there is recognition of the need to raise the standard of writing in humanities but uncertainty about how it can be done.

There are numerous criticisms that can be made of this approach. Three will be mentioned here. A practical one is that pupils need experience of writing for their examinations, especially the GCSE, with its wider educational objectives. A more general criticism is that writing should contribute to the learning process, giving the pupils the opportunity to speculate and discuss ideas. As in the writing of this chapter, it usually involves revision of first ideas before something acceptable is achieved. A third criticism is that the endless routine exercises involve little thinking, and pupils associate writing with boring work. What can happen is that they start to see writing as real work, while talking and reading are not. They settle into a routine where they are doing simple exercises, and, although they are not involved, they resist attempts by the teacher to introduce more challenging activities. With a difficult class the teacher may go back to the safe routines. Pupils and teacher collude in mediocrity.

ANALYSING THE PUPILS' WRITING: FIRST RESPONSES TO EXAMPLES OF EXTENDED WRITING

The pieces of writing below are examples of longer pieces of writing produced by Fatima and Ferdinand and provide the opportunity for a more detailed analysis of their writing skills. The background to this writing was a study of the home front during the Second World War. In the classroom the pupils read and discussed a booklet on the topic and were asked to write about the problems of people living in London at the time. Looking at their writing, my first concern is the intention of the writers and the meaning they are trying to convey. Ferdinand, for instance, seems to be fascinated, perhaps horrified, by the stories about bombing and focuses on this, excluding other issues. He presents a picture of relentless attacks and their effects on people's lives. Although the surface features are not well developed, and he refers to atom bombs that were not dropped on London, he is trying to express his ideas and shows empathy with the plight of people coping with the Blitz.

Writing in a humanities classroom

Bombs
The Trouble with Bombs were The noise of the guns The people could not sleep until the guns stop firing. when the guns starts firing The people would go into there shelters. They also have fire bombs which They drop everywhere and They were large bombs that They drop on houses and They also have flying bombs which they control Them to drop where they want them to drop The most powerful bomb of them all is a attom bomb which can distroy Britian with one attom bomb (*Ferdinand*)

The problems of the civilian population in Britain during the war
The British Civilian Population are the ordinary civilians who are not in the army, navy or air force. During the war Britain could only grow enough food for a third of its population and so there was a food shortage. The government decided to ration the food so everyone had an equal share. Only the heavy industry workers got extra rations, because they needed extra energy. Enemy ships blew up most of the ships that were bringing food to Britain, the civilians had to do without a lot of things. The Government encouraged people to eat home grown food and to dig up their flower gardens to grow vegetables. (*Fatima*)

Fatima, in contrast, has a different intention with her writing. She is aware of the demands of the teacher for a more dispassionate account that covers a range of issues. The rest of her piece describes rationing, the evacuees, the blackout, and gas masks, as well as air raids. After this descriptive account she expresses the opinion that 'People remember the soldiers who fought bravely in the war but they shouldn't forget the ordinary civilians who fought hard to keep up the home front.' She is able to step back and see a wider view of this difficult time. Generally she is a more accomplished writer who consciously tries to achieve the style wanted by the teacher.

ANALYSIS OF THE TEXTS

Looking at the grammar and style of the two writers in more detail, we can see how Ferdinand is at an earlier stage of writing development. Ferdinand's writing shows that he can use different kinds of sentence: simple, compound, and complex, with subordinate and relative clauses. They are written in a straightforward way without embedding, and the repetition of 'They + verb' from the fourth sentence, suggests limitations in his style. The verbs give some problems, e.g. 'the guns *starts*', 'the trouble with bombs *were*', and 'could not sleep until the guns *stop* firing'. He prefers to use the present tense and may be uncertain about the differences to do with number and person. Some verb elements are more complex, e.g. 'could not sleep', 'stop firing', and 'they control them to drop'. He uses a limited range of nouns, pronouns, and adjectives, the latter all for describing bombs. His presentation, not shown here, is readable but not carefully written. He often uses a capital 't' which coincides with a sentence starting 'They' or 'The', giving the false impression that he uses capitals to start a sentence. There are two full-stops, suggesting that he is unsure about punctuation. Spelling is mainly correct. Errors include 'there' instead of 'their', 'attom', 'Britian', and 'distroy'.

An analysis of Fatima's writing provides an interesting contrast. The first sentence defines the term 'civilian population' and she then writes a paragraph on food rationing. The meaning is clear and the sentences progress logically, using mainly a subject-verb-object framework with subordinate and relative clauses. Unlike Ferdinand, she uses more connectives such as 'so'. The last sentence shows effective use of embedded clauses, which contrasts with Ferdinand's repetitive 'and they'. The verbs are consistently in the past tense, with similar use of auxiliaries and participles but with more use of the infinitive. In the noun phrases there is a wider use of adjectives. The overall presentation is good, with careful handwriting, use of capitals at the start of sentences, reasonable punctuation, and correct spelling. Vague references to 'they' are not made at the start of the sentences and rarely within a sentence. She constantly refers to the subject in the noun form: 'Britain', 'the government', 'enemy ships', etc.

STAGE OF DEVELOPMENT

Table 2.2 sets out a simple and tentative division into earlier and later stages of writing development, outlining some of the features that can be looked for in the writing of secondary pupils. Comparing the examples of writing with the table, we can see that Ferdinand is at an earlier stage of development than Fatima, as suggested by the influence of speech patterns, the simple sentences, the dominance of the same sentence structure, and the repetition of the subject noun. He needs to develop the use of the past tense, embedded clauses, connectives, adjectives, capitals, and full stops, and to make explicit the subject or object of the sentence. He also needs to sequence sentences so that there is a logical progression. A qualification, however, to this analysis is that more examples of his work need to be examined. Looking through his writing from the first- and second-year humanities and from his English lessons, I have noticed that he can use features, such as embedding, which are found in later stages of writing.

Table 2.2 Earlier and later stages of writing

Earlier stage	Later stage (in some kinds of writing)
Simple sentence	Complex sentence with subordination and embedding
Connectives: 'and', 'when', 'then'	Connectives: 'so', 'because'
Active voice	Passive voice
Personal, using 'I' or 'we'	Impersonal, using the third person
Narrative with sequencing by time	Scientific genre with logical sequence
Present tense and regular past, e.g. with -ed endings	Irregular past tense, use of more auxiliaries and more complex verb forms
May assume teacher knows a lot of the background	Assumes text stands on its own and may be read by anyone
Repetition of same sentence structure and often the subject noun repeated in different sentences	Varied sentence structure and use of connective devices between sentences
Unsure about the use of capitals and full stops	Sentences accurately marked
Unsure about the use of apostrophe for possession and abbreviation	Correct use of apostrophe for possession and abbreviation
Little use of adjectives	Wide use of adjectives
Strong influence of speech on writing	Has taken on role of writing as different from speech

A more developed set of ideas can be found in Kress (1982). He takes the view that the writing of children is different from that of adults but not necessarily inferior. He studied the writing of 6-14 year olds, relating syntax and meaning, and argues that children often use a syntax appropriate to their cognitive stage. First of all he distinguishes between writing and speech. They differ in a number of ways, including text and syntax. Speech is based on the information unit in sequences of clauses, whereas adult writing is based on the theme and rheme in a hierarchic structure. Speech will influence the children when they start to write, but the difficulties involved may lead, initially, to the use of simpler sentences than in speech so as to concentrate on the meaning. A sentence of a child's writing is often related to the line, and is like a paragraph, with speech clauses linked textually and then syntactically. The clauses are connected by co-ordinating and adjoining, using words such as 'and', 'then', and 'when'. Kress says the child's view of causality is that of regularity, a sequence of events, and this means that narrative is a useful genre for motivating pupils towards adult writing. Based on a sequence of events, it can nevertheless have the writer as an observer or narrator, a distancing from the events of the story.

In addition Kress says that adult writing is hierarchical, with embedding and subordination. With the development of paragraphing more freedom is allowed within the sentence and there are more cohesive devices in and between sentences. The adult view of causality is that of power. Adults achieve connections using 'so' and 'because', and make more use of prepositions, such as 'with', 'by' and 'though'. The language is more impersonal and passive, as in factual genres which, in children, are often lists using 'to be' or 'to have', but which in adults are hierarchically structured and topically sequenced.

Ferdinand, using this framework, seems to be influenced by speech as he links speech clauses textually and syntactically, using co-ordination rather than embedding, with a 'regularity' view of causality more than a 'power' one, as shown in the use of 'and' and 'then' and 'when' rather than 'so' and 'because'. On the other hand, he does use embedding and subordination in some cases, so he has the ability to become more adult in his writing. Kress argues that children's writing should not be judged as adult language. Correction from an adult perspective is inappropriate and errors may be indicators of development. Underlying this view is the idea that the syntax and meaning of the child's writing are related. As a teacher I need to be aware of Ferdinand trying to sort out his ideas and express them from his point of view and his level of development. Comparison with adults or other children may not be helpful.

PERSONALITY AND LANGUAGE BACKGROUND: INDIVIDUAL DIFFERENCES

Ferdinand and Fatima are strongly contrasting individuals, and this difference may partly explain their varying success as writers. Fatima interacts a lot with the teacher and other children, asking and answering questions. She can concentrate, and works methodically through the task. She has a history of success, academically and socially, and brings it with her to the classroom. Looking at her exercise book, I noticed she had spelt 'their' as 'thier'. She told me she was following the rule 'i before e except after c'. I pointed out that this word did not follow the rule, and from then on she spelt it correctly. Her writing is neat, even, and clear. In her piece on the home front she made no false starts or errors that need crossing out. The sentences are logically connected and she has a clear sense of paragraphing. She has been able to read a booklet on the topic and select information, putting it together in her own way. She knows what the teachers want, and she likes to follow the rules and have a disciplined environment. If the teacher asks the class to be quiet, she stops talking straight away and listens. Words that come to mind to describe her personality are 'able', 'ordered', 'logical', 'precise', 'sensitive', 'assertive', and 'helpful'.

Compare this with Ferdinand. His greatest strength is his willingness to try, but he feels very unsure sometimes about what he has to do. It is typical of him not to finish the first set of questions in the humanities lesson and to go to the next set. He is concerned with the content of his work but is less concerned with the form. Perhaps he puts so much energy into sorting out the meaning that the surface features are neglected. His English teacher has observed his writing regressing when he is trying out new ideas. Certainly in his writing there is a sense of uncertainty. Although he is always agreeable, to some extent passive, he clearly has anxieties, as shown in his desire to keep his work at home, the way he hides his work in class, and his unwillingness to ask a teacher for help. A description of him might be: 'conscientious, passive, amenable, confused, and lacking in confidence'.

The idea that personality can influence writing development may seem straightforward but does, in fact, raise

Table 2.3 Language and personality, based on the ideas of Jung (1964)

Characteristic	Description	Language
Thinking	Tells you what it is	Ideational, referential, logical, impersonal
Feeling	Tells you whether it is agreeable	Judging, evaluating, preferential, personal
Sensation	Tells you something exists	Factual, descriptive, sensory
Intuition	Tells you whence it came	Imaginative, artistic, creative
Introversion	Inward, thoughtful	Reflective, personal, private
Extroversion	Outward, takes risks	Expressive, public

a number of issues that are not easily resolved. One difficulty arises where the personality is seen as a fixed attribute, usually the result of family background and individual characteristics. There needs to be recognition, as well, that personality can change with time and that it can be situational, varying, say, between school and home. Also, judging personality from our observations and experience may mean lack of objectivity. An alternative is to use personality theory to help us, but there are many competing theories (Fontana, 1977). Table 2.3 takes one of these and relates personality characteristics to language. Personalities can be located on three continua, thinking-feeling, sensation-intuition, and introversion-extroversion, and actual placing on the continua would possibly lead to a predisposition towards certain styles of language. This in turn might be linked to cultural or gender differences in society. In Brice-Heath's (1983) study of two communities in America, Roadville and Tracton, there is a contrast between language development, child rearing, and personal relationships. Tracton children have to learn to assert themselves, breaking in to adult discourse. In Roadville the children must pay attention, listen, and behave, with the emphasis on information rather than imagination. This kind of study explores the complex relationship between culture and language without stereotyping the link between culture and personality. The general implication is that a deeper understanding of personality can help us understand the writing development of individual children but there are many difficulties involved.

LANGUAGE BACKGROUND

Ferdinand is one of those pupils who speak mainly their mother tongue at home. Occasionally he uses English with his parents and sometimes with his cousins. He came to Britain about five years ago, so he has had to face some of the difficulties of new arrivals, although he had some experience of English in the Philippines, where it is the official language. In the last year of the junior school his reading age was 7.1 and in the second year of his next school he scored 8.8, about four years below his real age. He has had special lessons with ESL teachers, mainly in his first year at secondary school. At the end of that year the ESL teacher reported that his discussion and oral work was good, and he could find meaning in complicated passages, but he had difficulty with writing. In contrast, Fatima was born in Britain and speaks English at home with her parents. She picked up some of the Mauritian Creole from home but learnt it more thoroughly when she visited Mauritius. She is the only student to have a reading age above her chronological age.

There is controversy about the effect of bilingualism on English development which I also address more generally in a subsequent chapter. One view is that bilingualism should be encouraged as a positive quality and that, in the right circumstances, it can be an asset for English language development. It is pointed out that bilingualism is common in many parts of the world and that children can learn to be fluent in two or more languages. Other things being equal, being bilingual would not affect the pupil's level of achievement in school. You would find successful pupils like Fatima, and less successful ones like Ferdinand, both bilingual but varying in their achievement. This is supported by the measurements of reading age in their class. Table 2.4 shows the results of a London Reading Test given to the class halfway through their second year. Bearing in mind the limitations of these standardized tests, the results range from 8.8 to 13.5, with an average of 10.4, compared with the average chronological age of 12 years 11 months. Two recent arrivals had no score on the test and are not included. The table shows bilingual, Creole-influenced, and monolingual children distributed evenly.

Table 2.4 Reading age scores and language background

Reading age score	Language background
13.5	Bilingual (Mauritian Creole). Mainly English at home
12.2	London towards standard
12.2	London towards Cockney
11.0	London. Creole influences from home and friends
10.8	Bilingual (Greek Cypriot). English encouraged at home
10.7	London. Creole influences from home and friends
10.7	Bilingual (Gujerati). English encouraged at home
10.5	London. Creole influences from home and friends
10.3	Bilingual (Turkish Cypriot). English encouraged at home
9.9	London towards Cockney
9.8	Bilingual (Greek Cypriot). Mainly Greek at home
9.7	Bilingual (Greek Cypriot). Mainly Greek at home
9.6	London. Creole influences from home and friends
9.6	Bilingual (Greek Cypriot). Mainly Greek at home
9.0	Bilingual (Urdu). Mainly Urdu at home
8.8	Bilingual (Filipino). Mainly Filipino at home
8.8	London/London Jamaican

The London Reading Test was taken on 26 January 1984. The average reading age is 10.4. The average chronological age is 12 years 11 months, ranging from 12.2 to 13.3. Children from a 'Caribbean' background shift the way they talk from situation to situation, influenced by London, Creole, and London Jamaican dialects. Bilinguals have no more problems than other students.

The order of bilingual children, however, seems to be related to the proportion of English used at home. Those who speak mainly English at home, and whose speech is nearer to standard, are most successful, and those who speak mainly the mother tongue at home are least successful. Fatima and Ferdinand illustrate the difference. Allowing for the limitations of one case study, this supports the view that some bilingual pupils can be held back if they have not had enough time to practise their English. Their lack of progress may also be influenced by other factors. It may be racism that restricts the children's access to the wider community. Then there is the difference between literacy and speech. Competence at talking might be satisfactory but the home may be less supportive to the development of literacy, in either language. The child's history at school may have had an effect through inadequate or even racist reading schemes, rather than a supportive bilingual policy. The fact that some children with less experience of English do less well could also be seen as an indicator of the inadequacy of school in dealing with their needs. Finally, there will be individual differences between pupils, some able to do well almost regardless of circumstances, other always needing extra help.

SOCIAL FACTORS

Gender, class, and culture have an important effect on society and education, and the response of schools may perpetuate or create differential treatment that can cause inequality in educational provision and achievement. Language, including writing, will be similarly affected, and the relationship between language and social structure has been the subject of a wide range of studies. Two introductory books that I have found useful are Trudgill (1974) and Stubbs (1976). As I see it, they argue that most children, regardless of their social group, have a basic linguistic competence, and although there may be differences in the style of language used by different groups, working-class and black children do *not* suffer from language deprivation. If these children are less successful at school, then it is due to social and educational factors. Poor social conditions may disadvantage children because of their lack of access, say, to the books and styles of language needed in education. Within schools the negative attitudes of teachers to the home language and culture, and the methods of teaching, may further disadvantage these children. Difficulties in writing are a reflection of social and educational disadvantage, not linguistic deprivation.

From this analysis the writing demands of school will create more difficulties for black and working-class children because they are furthest removed from the language *and* culture of home. The different styles of language used by different groups creates differences in preference and experience. Formal writing is closer to the styles of middle-class white children, so the shift to writing involves less of a language *and* cultural shift for those pupils. Black and working-class children have the ability to be competent writers but need language teaching appropriate to their needs. This effect will be more pronounced with scientific and impersonal writing, which may also disadvantage girls in school. The argument is that impersonal, technical, and abstract language is associated with a scientific approach which, in our culture and schools, attracts the interest of white middle-class boys but may alienate many girls and black and working-class children. The latter groups have the ability to use a scientific genre of writing and become competent in science, but are less willing to do so because of the cultural associations of the subject and the process of socialization they have encountered in society and in school.

These general ideas about language and social structure are difficult to apply to particular people, such as Fatima and Ferdinand. One problem is that they are oversimplified and ignore many of the complexities and controversies to do with this issue. A second point is that, although social factors have a general effect, particular people and families may not illustrate these tendencies. Class, gender, and culture will affect the personalities and language background of the two pupils, as already described, but there will also be an individual component, linked to their own particular circumstances and history. Fatima, for instance, does not fit the pattern described above. She is female and working-class, and from a religious and ethnic minority, but likes scientific work and enjoys the use of formal language styles. Her personality and family background have been strong factors, and her experience in school has been positive. Ferdinand seems to fit the pattern, to some extent, but his low attainment might equally be due to individual circumstances.

Social factors may be more apparent when studying the whole class. Generally the tutor group is multicultural and working-class, with a low level of attainment in English, as indicated by the scores for reading age (table 2.4). As a geography and humanities teacher my concern for their poor writing skills was one of the reasons why I decided to find out more about language development. As I see it, the children have linguistic competence but may lack the experience of styles of language used in school, especially those to do with literacy. The demand by schools for all children to use standard English and accepted writing styles, or genres, is legitimate. Schools have the responsibility to value the language and culture of home and, at the same time, extend the children towards these less familiar language registers. For the children in this school, the development of a curriculum that takes on the issues of gender, class, and culture is a necessary condition for this language development. Humanities education is particularly valuable in this area because it can provide a supportive curriculum for this work, with its wide range of content, skills, and language activities. For writing, this would include the suggestions found on p.42. Social factors, then, can be seen, to some extent, in the general writing needs of a class or school, but when we consider individuals we need to know about their own personal background and circumstances, as can be seen in the studies of Fatima and Ferdinand.

CONCLUSION

This case study has applied some of the general ideas about writing development to the work of two pupils studying humanities. One set of ideas encourages a wide range of writing activities and an open process which allows opportunities for drafting and revision and collaborative work. In the school I studied this was not well developed, although some extended writing was set, such as the work on the Blitz. Teachers are resistant, hesitant, and sceptical about the value of this approach and need convincing that it will help the pupils learn the necessary concepts and skills. A second set of ideas asks teachers to mark pupils' work individually, according to their stage of development and particular needs. In this example it is Ferdinand who is having difficulties with his writing and needs special consideration. An initial response to his writing might be that it is inadequate in its content, grammar, and presentation. An alternative view is to see it as his attempt to make sense of the task given to him, as in his writing on the bombing.

A third set of ideas asks us to consider the personality, language, and social background of each pupil. The danger of this approach is that the pupils may be perceived as deficient, because of their family circumstances or personality, and this may lead to low expectations from the teacher and poor levels of achievement. This may come about through a negative assessment of the behaviour, attitude, and ability of the pupil and prejudice towards his or her class, gender, and culture. I take the view that the work done in school can make a significant contribution to writing development and would oppose fatalistic attitudes that suggest schools have little influence. Although it is difficult to detail the significance of social factors and the way schools respond to them, their importance needs to be recognized, and schools need to ensure that they are providing a climate of learning that will help all children to write, including children such as Fatima and Ferdinand.

REFERENCES

Brice-Heath, S. (1983) *Ways with Words*, Cambridge: Cambridge University Press.
Britton, J., et al. (1975) *The Development of Writing Abilities (11-18)*, London: Macmillan.
Czerniewska, P. (1981) *Describing Language: an Introduction to Grammar*, Milton Keynes: Open University.
Fontana, D. (1977) *Teaching and Personality*, Oxford: Blackwell.
Jung, C. (1964) *Man and his Symbols*, London: Aldus.
Kress, G. (1982) *Learning to Write*, London: Routledge & Kegan Paul.
Martin, N., D'Arcy, P., Newton, B., Parker, R. (1976) *Writing and Learning across the Curriculum, 11-16*, London: Ward Lock.
Richmond, J. (1986) 'What we need when we write', *About Writing*, Newsletter No. 2 of the National Writing Project, London: SCDC
Stubbs, M. (1976) *Language, Schools and Classrooms*, London: Methuen.
Trudgill, P. (1974) *Sociolinguistics*, Harmondsworth: Penguin.

Chapter Three

A CASE STUDY OF LANGUAGE AND LEARNING IN PHYSICAL GEOGRAPHY

Sharon Hamilton-Wieler

> As a geography teacher, I find I am a teacher of reading and a teacher of writing ... (P. McLeod, 1985)

Meet Peter McLeod, head of the geography department in Crown Woods School in south London. Because of his interest in language as a means by which geographical concepts are organized and articulated, he offered to participate in a cross-curricular investigation of the language environments of A-level classrooms, and how they influence the nature of the writing and the nature of the learning which occur in these classrooms. (1) From September 1984 until July 1985 I observed Peter McLeod and his twelve upper sixth-form students preparing for the A-level examination in physical geography. Although preparation for demonstrating acquired information and knowledge in an examination shaped and influenced most of the teaching-learning interactions which occurred during that year, to foreground examination performance as the sole or even primary function of that classroom would be to overlook some major features of the manner in which McLeod and his students engaged with the body of knowledge which comprises physical geography.

BASIC FUNCTIONS OF A-LEVEL CLASSROOMS

The basic functions common to all A-level classrooms could be categorized as follows:

1 To develop and extend students' specialized knowledge and understanding of the discipline.
2 To prepare students to demonstrate to an unknown examiner their ability to recall discipline-specific concepts and principles, to apply this knowledge to novel situations, and to articulate this application of knowledge in

coherently organized written text, using appropriate discipline-specific terminology competently and confidently.

These two functions have the potential to work at cross-purposes, placing both teacher and students in a fundamental dilemma which positions writing as the site of a tremendous struggle in the A-level context. Discussions of the relationship between language and learning which have proliferated over the past twenty years emphasize that the nature of task, text, and context which encourages learning and understanding is significantly different from the nature of task, text, and context which requires demonstrating to an unknown examiner acquired information and knowledge. (2)

FRAMING THE QUESTION

The resultant dilemma can be framed by posing the following question: (*a*) within the constraints of time and a content-laden syllabus, and (*b*) in an educational context wherein language functions predominantly as a means of preparing students to demonstrate knowledge in writing on future-determining examinations, <u>how</u> can teachers provide opportunities for their students to use language - both spoken and written - in its heuristic capacity to explore discipline-specific concepts, thereby deepening and extending their understanding of the new bodies of knowledge which are confronting them?

That McLeod's classroom offers these opportunities can be inferred from his students' attitudes towards writing about geography. While recognizing that different students have differing perceptions of what occurs in any particular classroom, and that one student's viewpoint will not necessarily represent the collective response of his or her classmates, none the less, based on my own year-long observation of student-teacher interactions in McLeod's classroom, I offer these comments made by Nigel, during one of our talks about writing, as a means of entering the language environment created by McLeod during his sixth-form physical geography lessons:

> Out of my A-level subjects [biology, mathematics, geography] writing is undoubtedly the most important in geography ... writing in geography helps you to argue things better ... there's no certainty about anything in geomorphology, so you can't really be wrong. There's lots of opportunity to speculate ... the best part of writing in

A case study: physical geography

geography is writing about your fieldwork ... it's your own analysis ... you set the whole thing up yourself ... you're left to think for yourself ... Geography is absolutely massive in scope. You can always carry your ideas further. (Nigel, 21 March 1985)

Nigel's comments highlight several issues relevant to creating a classroom context which is sensitive to relationships between language and learning. His overarching thesis that 'writing in geography helps you to argue things better' is elaborated in the statements which follow, statements relating directly to the manner in which McLeod encourages his students to engage with geographical concepts in relation to the observable world in which they live.

'THERE'S NO CERTAINTY'

For example, the confident assertion that 'there's no certainty about anything in geomorphology, so you can't really be wrong' emerges from McLeod's insistence that students be aware of the controversies and differing opinions in the literature about land formations, and be alert in their own observations in order to develop some means of critically assessing opposing viewpoints. This demand assumes a familiarity with the literature, an assumption which returns us to the words I used to introduce McLeod at the outset: 'As a geography teacher, I find I am a teacher of reading and a teacher of writing ...' Reading was an essential component of the talking and writing which occurred in this classroom throughout the period of my observation. It was not unusual for each student to arrive with three or four books in hand, related to the day's topic. Nor was it unusual for McLeod to have a number of books readily available in the classroom, as well as references to books or to chapters of books available in the school or community library. These book and chapter listings highlighted differing opinions or positions about geomorphological formations and processes of change. Presentation of new material and classroom discussions focused on these differing viewpoints, exploring the dissonances and, when appropriate, relating them to students' own observations during their field experiences. Books and students' field notes were kept close at hand for consultation or support when controversies flared. What resulted was informed and lively discussions - or arguments - wherein arriving at conclusions was secondary to the primary goal of exploring alternatives. This dialogic feature of classroom discussions was followed through as an expectation in pupils' written work. During one of our frequent conversations about writing in geography, McLeod stated:

> A lot of thinking about land forms is conjecture and hypothesis ... there is always room for more research, alternative explanations, doubts, interpretations ... I'm a great believer in 'maybe' and 'possibly' and 'implies' ... their essays should not simply be an assemblage of facts ... there should be some realization of controversy and disagreement. Even if concrete research has been quoted, one might look for words and phrases like 'however', 'perhaps', 'maybe', 'recent research suggests', 'some would argue that', 'on the other hand', 'possibly' ... (26 March 1985)

It is evident from Nigel's comments that McLeod's emphasis on exploring alternatives has freed him from the 'right or wrong' view of written responses which is so pervasive in the educational context. Instead of the irksome, at times almost insurmountable, constraint of needing to converge upon 'the correct answer', Nigel obviously enjoys the opportunity to play with a range of ideas in relation to his observable world.

'OPPORTUNITY TO SPECULATE'

Nigel's next statement, 'There's lots of opportunity to speculate', while closely related to the preceding discussion, points out another characteristic feature of McLeod's classroom. Discussions of differing viewpoints presented by expert geomorphologists, essential to shaping students' awareness of ways of perceiving geographical phenomena, none the less vest the balance of power - of authority - in the printed word. But speculation, particularly informed speculation which integrates students' knowledge acquired from reading and from lectures with their experiences in the world outside the school classroom, has the potential to vest a significant degree of authority in the student. That students did not readily take up this opportunity at the beginning of the year was a source of concern to McLeod:

> I find it disconcerting. I try to get them to refer to their field experience, and when it supports what they've read, they do. But if there seems to be a difference between what they've observed and what they've read, they too

often dismiss their own conclusions or recorded observations as invalid or wrong, and refer only to the text. (October 1984)

McLeod persisted in encouraging his students throughout the year to confront the printed word with the evidence of their own observations, with the result that, after their final major experience in the field, they did begin to incorporate more of their fieldwork into their written and oral discussion. Even so, their attempts were sporadic and tentative, possibly suggesting that the pedagogical assumptions behind several years of institutional validation of printed text over personal experience are not easily questioned or overcome.

'WRITING ABOUT YOUR OWN FIELDWORK'

It is the fieldwork component of physical geography, particularly the individual study, which provides students with their major opportunity to integrate their knowledge and understanding of geographical concepts acquired through reading and lectures with their own on-site observations, reflections, and conclusions. Since students select their own problem or topic, organize and carry out their own fieldwork, and decide on the written format of presentation most appropriate to their chosen study, they are indeed taking upon themselves the role of protogeomorphologists or apprentice geographers. As Nigel tells us, 'the best part of writing in geography is writing about your own fieldwork ... it's your own analysis ... you set the whole thing up yourself ... you're left more to think for yourself.'

I mentioned that the dual functions of the A-level classroom have the potential to position writing as the site of a struggle, a struggle which could be characterized as a tension between the institutional expectations of competent academic prose and personal engagement with discipline-specific evidence. Harold Rosen, Emeritus Professor of the Institute of Education, University of London, tells the story of one of his two sons who is today a geologist of international reputation. When asked in the fourth form to define a particular geological formation, he began his definition with the words 'an exciting projection of rock which ...'. The teacher wrote beside his response, 'Whether or not it is exciting is not important.' I am reminded of Rosen's story when I read Nigel's independent study. It is similar to the shorter teacher-assigned writing tasks he has done, in that it gives evidence of a depth of understanding based on background reading and reveals Nigel as using this reading to raise his own questions, once again a writing strategy - and conceptual strategy - emphasized and encouraged by McLeod. Where the study differs dramatically from his other texts is that his questions arise out of problems in his own investigations, in his own experience, and so his writing is concerned with first-hand engagement with geographical phenomena. What I find surprising in Nigel's study is that despite this first-hand experience, and despite the evident excitement and enthusiasm he displayed when talking to me about his study, the resultant text has submerged his strong personal engagement with the question he was investigating, and is expressed in the 'agentless' prose of what McLeod calls 'good, objective, academic writing':

> This study aims to investigate the properties of a straight channel including the nature of water flow, the occurrence of riffles and pools, the bedload etc. and to compare these with those of meandering channels. The findings will be presented diagrammatically and compared with those of researchers such as Leopold and Langbein (1966). It should then be possible to account for the occurrence of straight channels in nature. (25 February 1985)

'I' ENTERS THE TEXT

There are two exceptions wherein Nigel allows the usually suppressed 'I' to enter his text, the first in the 'Methods and problems encountered' section:

> The worst problem came when using the data to plot a depth and velocity map of the reach. The most detailed plan available (by courtesy of the GLC) was a scale of 1:1250, and this plan was obviously too small to be used for showing the measurements. Eventually I decided to scale this map up until it was sufficiently large to use to show channel depth and velocity. It might have been better to do a compass traverse survey of the river and plot the channel using the bearings but this would have taken too long. I decided that, in any case, the scale at which a compass survey map could practically be drawn would soon iron out any minor irregularities identified, so that to all intents and purposes it would be little improvement on the enlarged version. Field measurements were made sufficiently detailed so that errors in the outline map could be corrected. Particular features of the channel such as

A case study: physical geography

undercutting, obstacles and deposition banks were also noted and measured for inclusion on the maps.

I find this example interesting because it shows Nigel, the researcher, trying to sort out a problem and having to settle on a less than satisfactory solution. Once he has described and rationalized his decision to the point where he, in actually working through the decision-making process, feels comfortable that it was an acceptable solution to his problem, he reverts immediately to the agentless passive: 'Field measurements were made ...', 'Particular features of the channel ... were noted'.

The second exception occurs in the 'Conclusions and discussion' section, wherein a part of his original hypothesis is not supported by his data in quite the way he anticipated. In searching for an explanation he worked through the data to a moment of insight; in describing this moment of insight he puts himself, syntactically, into his text:

> In this study it was originally thought that angular pebbles would assist in the accumulation of material by trapping and interlocking with rounder particles and thus promote the development of shallowing associated with riffles. These would therefore be likely to contain a higher proportion of angular particles and fewer well rounded pebbles compared with the bedload of pools.
>
> What emerged from the overall statistics was surprising, riffles did contain, on average, more angular particles and less well rounded particles but at the same time pools contained much fewer 'rounded' pebbles and more 'sub-angular' pebbles while the proportion of sub-rounded particles was relatively similar. It could be that the effect of the 'sub-rounded' pebbles in a pool is equivalent to that of the 'angular' pebbles or the less 'rounded' pebbles in a riffle.
>
> I then realized, however, that if angular particles lead to greater stability of the bed then it would actually be destructive to the river's established pattern of energy loss if there was a difference in stability between riffles and pools. If the bedload of pools, for example, was less stable they might 'migrate' downstream more readily and encroach on the following riffle, thus distorting the balance in energy loss conferred by the alternative sequence. It seems imperative therefore that one form should be no more or less stable than the other. The reason for the observed differences in roundness of the bedload samples then becomes clear.

A moment of decision and a moment of realization, both reflected syntactically in the text, both aberrations from Nigel's customary objective prose style; three two-word phrases in 4,150 words: 'I decided', 'I decided', 'I realized'. Because of their 'exceptional' nature they seem important; they seem to signal learning of a different nature occurring. The situation is more than the customary application of facts or synthesis of information. Nigel has chosen a question to investigate, a question which leads not only out of the research of other geographers but also out of his own interest in a brook which cuts behind his own back garden. And in attempting to answer that question through his own investigative fieldwork he has taken up the role of geographer. This is a vastly different role from his usual student-in-the-classroom role, in which he primarily writes about the theories or discoveries or investigative procedures of others. I wonder whether, when composing this paper, Nigel was reliving the difficulty that he, in the role of geographer, experienced in making a decision which seemed to him less than satisfactory, and was reliving the excitement that he, again in the role of geographer, experienced when, through his own mental effort, a natural phenomenon which had puzzled him became clear. I wonder, is it because these two cognitive acts, 'I decided' and 'I realized', were acts of Nigel-the-geographer rather than of Nigel-the-student that he grants them this 'exceptional' subjective status in his paper? Whatever the answer, it is evident that the individual study functions to put Nigel, and his fellow students, into a first-hand engagement with geographical issues and evidence, and therefore with the discourse of geography rather than with academic discourse about geography.

WRITING IN THE CLASSROOM

Yet the reading, writing, and critical thinking abilities essential to designing and carrying out a study such as Nigel's must be, at least in part, influenced by the day-to-day explorations and discussions of geographical concepts which occur in the classroom. Since I cannot summarize in these few pages all that occurred during my year of observations in McLeod's classroom, I will focus on one particular writing activity to try to illustrate how a teacher's view of relationships between language and learning exerts a critical influence on his students' writing and learning.

After reading a class set of essay responses to the question 'What processes have shaped valley-side slopes in humid-temperature areas?' McLeod became concerned that his students were experiencing difficulty in determining not what specific points to include but what *kind* of points to bring to bear in response to a particular task. He therefore planned the following activity to try to enable his students to develop a procedure - a set of strategies - to help them respond to the informational and conceptual requirement of a particular writing task. After writing the topic 'Why do rates of marine erosion vary from time to time and from place to place?' on the board, he suggested the following process as a way into responding to the question:

1. Individually students brainstormed on paper, putting down any words, phrases, or lines of argument which came to mind (about three to five minutes)
2. In groups of three, students sorted through their collective responses and categorized them hierarchically - main points, supporting details, related ideas, etc. Students were encouraged to use their star diagrams, mapping and webbing techniques, and outlines - whatever seemed most appropriate (about ten minutes).
3. Time was provided to verify and amplify their information from texts, reference books, and lecture notes (about fifteen minutes).
4. Each group collaborated to generate one outline of the line of argument - with supporting details - it would pursue in developing a response to the writing task (twenty to twenty-five minutes).
5. Each group's outline was written on acetate and projected on to a screen for the class as a whole to analyse and discuss (about twenty-five minutes).

I chose to highlight this particular writing activity because it is embedded in traditional pedagogy: the writing task is teacher-constructed for the teacher's purposes, and assigned to the whole class, apparently assuming the existence or possibility of a correct or somewhat convergent response. Yet the context - the process of responding - set up by McLeod is tremendously sensitive to current understanding of relationships between language and learning. The initial brainstorming activity encourages each student to use processes of association to tap his or her recallable and tacit knowledge. Most students were pleasantly surprised to discover how much specific geographical knowledge came readily to mind, and how much of their broader tacit knowledge of the world they could draw upon in relation to the question. The group sorting activity requires processes of selection, categorization, and more association as they talk through the task. The following fragments of discussion indicate the complexity of decision-making and degree of collaboration required to accomplish this task:

> Would you start with ...
> Have you considered ...
> What do you think about ...
> Next, I think we should bring in ...
> Do you think we ought to mention ...
> Would we give a whole paragraph to ... or would we just ...
> Let's see - we have to show whether it's a destructive or a constructive force ...
> Yes, we did that ...
> Now, the question is: how does it vary from time to time and from place to place ...
> Is it just the direction, or do we need to include ...
> How do we bring in ...
> We should put about the teachers ...
> Did we mention ...
> What's it called when ...
> It's a bit like a [unintelligible], isn't it? The coastline? ...
> The coastline - yeah ...
> Don't you put in these bits? ...
> It's the transportation, the deposition, how it varies ...
> Did you do any rate measurement? ... How it varies ...
> We'll rough out the ...
> Yes, put that in ...
> Would you conclude that with ...

Verifying with printed material as an intermediate rather than beginning step puts the responsibility for factual information, initial conceptualization, and formulation of the response with the students. It forestalls the likelihood of a response organized similarly to the organization of a textbook or teacher notes. It also provides the students with a no-risk opportunity to assess and then extend their range of knowledge on a particular topic. Informational gaps can readily be filled in without loss of self-esteem. The next step, generating a line of argument as a group, allows a rich language-based opportunity to explore and discuss a variety of ways of organizing a response, and to develop some means of critically assessing why one might be preferable to another. These skills are further developed in the final step of the process, during which students examine and discuss each group's proposed line of argument. In just under eighty

A case study: physical geography

minutes these students engaged in a tremendous amount of talking, reading, writing, and critical thinking about geographical concepts as they worked collaboratively to negotiate a meaningful and effective response to a particular writing problem. Moreover they were involved in discovering ways of determining for themselves what constitutes 'meaningfulness' and 'effectiveness' by generating, examining, and assessing a number of alternatives.

A DESCRIPTIVE CASE STUDY

This case study is offered not as a prescription of what to do when teaching geography but as a description of what one teacher did with one class, and how one student responded to his teacher's instruction. As part of a larger study addressing the question 'How does writing emerge from the classroom context?' it attempts to show how critical is the teacher's role in enabling students to transform information, knowledge, and understanding to written text which enters the discourse of the discipline competently and confidently. It shows writing to be a mode of inquiry which embraces a range of language transactions and offers an invitation to further learning.

NOTES

1 Hamilton-Wieler, S. (1986).
2 For more detailed discussions of interrelationships between language and learning, particularly in the educational context, the following references should be helpful.

REFERENCES

Barnes, D., Britton, J., and Rosen, H. (1971) *Language, the Learner and the School*, 2nd edn, Harmondsworth: Penguin.
Britton, J. (1972) *Language and Learning*, Harmondsworth: Penguin.
Britton, J., et al. (1975) *The Development of Writing Abilities, 11-18*, London: Macmillan.
Bruner, J.S. (1975) 'Language as an instrument of thought', in A. Davies (ed.), *Problems of Language and Learning*, London: Heinemann.
Faigley, L., and Hansen, K. (1985) 'Learning to write in the social sciences', *College Composition and Communication*

36: 140-9.
Hamilton-Wieler, S. (1986) 'A context-based study of the writing of eighteen-year-olds, with special reference to A-level Biology, English, Geography, History, History of Art and Sociology', unpublished PhD thesis, University of London Institute of Education.
Martin, N., Medway, P., and Smith, H. (1973) *From Information to Understanding: Writing across the Curriculum, 11-13 years*, University of London, Institute of Education, and the Schools Council.
Martin, N., D'Arcy, P., Newton, B., and Parker, R. (1976) *Writing and Learning*, London: Ward Lock.
Medway, P. (1973) *From Talking to Writing: Writing across the Curriculum, 11-13 years*, University of London, Institute of Education, and the Schools Council.

Chapter Four

TALKING AND PROBLEM-SOLVING: REALITY ORIENTATED PROBLEM SOLVING QUESTIONS: A STRATEGY FOR INDUCING PRODUCTIVE VERBALIZATION FOR EXTERNALIZING AND CORRECTING MISCONCEPTIONS IN MAP WORK IN SECONDARY SCHOOLS

Julie I.N. Okpala

INTRODUCTION

The purpose of this study was to distinguish the extent to which misconceptions in map work would be discovered and corrected during classroom verbal interaction in two teaching situations. The teaching was in preparation for answering the West African Examinations Council (WAEC) type of questions in map work in school certificate examinations on the one hand, and the Reality Orientated Problem Solving (ROPS) question type on the other hand. Cohorts of school certificate geography students were taught for six weeks each in the two different teaching styles. Classroom verbal interactions were tape-recorded. Information elicited from the interaction showed that misconceptions were significantly concealed in WAEC lessons. ROPS lessons did not only significantly facilitate externalization of misconceptions but significantly showed evidence of attempts at correcting them. Some misconceptions observed in the ROPS lesson only form the crux of students' poor performance in the map-work examination (lack of use of the key and lack of understanding of word meaning of contour).

PRODUCTIVE VERBALIZATION

Language plays an important part in learning and communication when there is productive verbalization. Productive verbalization means that the inner thought of the learner is externalized. Piaget (1959), Lunnon (1969), and Ghaye (1986) are some researchers whose work on language and concept formation has been useful in drawing attention to the role of language and the use of language in development of concepts. The Language across the Curriculum project group emphasized that, for concepts to be developed, talking should not only be encouraged but should be expressive (exploratory).

MAP WORK

There is ample evidence that gaps and inconsistencies exist in students' learning of map work in secondary schools, hence academic performance in map work in school certificate examinations is poor (WAEC, 1975-84; Boardman, 1983). This literature reports that students find the location and interpretation of topographical features difficult. Geographers and examining bodies such as the WAEC (1975-84) have emphasized the need for improving strategies used in teaching map work in schools. Okpala (1987) in considering strategies for improving the teaching of geography agreed with educationists (Macintosh, 1970; Owen, 1973; Marsden, 1976) who propound that the restructuring of teaching and learning of any subject should start with restructuring of the pattern of questioning in that area in external examinations.

Map-work questions in West African school certificate demand theoretical understanding of concepts in the area. The questions demand identification of topographical features, shading specific heights, performing some mathematical calculations such as distance, gradient, and describing human and physical features. An example of such a question is given below.

(a) Draw a square to the scale of 1:50,000, to represent the outline of the area shown on the map. On this outline, mark and label:

(i) longitude 11°05'E and latitude 7°21'N.
(ii) the full length of the River Selbe and an arrow to indicate the direction of flow.
(iii) the named right bank tributary of the River Selbe.
(iv) the height of the trigonometrical station at Kwenta Walu.

(b) (i) Calculate the distance in kilometres as the crow flies from Mayo Selbe Rest House to the trigonometrical station at Kwenta (Walu).
(ii) State the bearing of May Selbe Rest House from the trigonometrical station at Kwenta (Walu).

(iii) Imagine that you are travelling in a car along the secondary road from the south-west to the north-east. Describe the scenery on both sides of the road using only the evidence of the map. (WAEC, 1982)

Okpala (1987) argued that the above pattern of questioning is theoretical and therefore its teaching is bound to be conventional and will not facilitate conceptualization. She therefore proposed Reality Orientated Problem Solving questions for WAEC map-work examinations. In ROPS questions realistic issues and problems which require the use of concepts in map work for their solution are presented to the examinee. He or she then culls up essential concepts and ideas for their solution. Such problems could be structured on incidents such as laying of pipes, locating a television mast, choice of route, construction of a new road or location of an industry. An example of such a question is given below.

The government of Nigeria has decided to build a road linking the Niels Valley area to the major road which runs north to south through Naraguta Hausawa and Jos.
The people living in Neils Valley would like the route to extend straight into the valley due east from Naraguta Hausawa.
You are the engineer in charge of advising the government on new routes. You feel that the direct link from Naraguta Hausawa is not possible.

(i) Write out your brief explanation (based on evidence from the map) to the people in Neils Valley why this direct link is not possible.
(ii) Briefly describe the route which you would recommend and give your reasons. (Use sketches where necessary.) (Okpala, 1987: 511-12)

The ROPS question above demands identification of features, calculation of distances and gradient and/or drawing of a cross section for its solution. A question which arises is: will setting ROPS questions in school certificate examinations facilitate teaching for understanding of concepts in map work?

STATEMENT OF THE PROBLEM

The problem investigated in this study was whether setting questions of the ROPS type will induce productive verbalization that would facilitate discovering and correcting misconceptions in map-work lessons.

Research questions

In order to tackle the problem the following research questions were asked:

1. Will the proportion of units of thought verbalized by students in ROPS lessons be higher than in the WAEC lessons?
2. Which lesson type, WAEC or ROPS, will have a higher frequency of lessons with incidences of misconceptions discovered and corrected?
3. Will the incidence of the types of misconceptions differ by lesson type - WAEC or ROPS?

Hypothesis

In order to make decisions on the results, the following hypotheses were formulated:

1. The proportions of units of thought by students will be significantly higher in the ROPS than in the WAEC teaching.
2. ROPS teaching will have a significantly higher proportion of lessons with misconceptions discovered and corrected than the WAEC.
3. ROPS teaching will show up a significantly higher proportion of misconceptions in the lessons as well as a higher proportion of lessons with other types of misconceptions than the WAEC.

METHODOLOGY

The design of this study was quasi-experimental. Eight teachers in eight randomly selected schools in Lagos, Nigeria, formed a sample for the study. The teachers taught map work for six weeks to cohorts of school certificate geography students in 1983/4 for the WAEC type and in 1984/5 for the ROPS question type. Thirty lessons were taught for a total of 1,082 minutes for the WAEC and twenty-eight lessons for 1,141 minutes on the ROPS question type.

To acquaint them with the ROPS question type, teachers were presented with two sets of equivalent questions on map work for the WAEC and ROPS. The equivalence of the concepts was confirmed by experts. The relationship between performance in the sets of the WAEC and the ROPS questions administered to a sample of 211 school certificate students were also found to be positive and significant beyond the 0.05 confidence level ($r = 0.58$ to 0.67). The correlation was also found to be significant by school types and by sex ($r = 0.35$ to 0.71).

In teaching for the WAEC question type, the normal teaching of map work by the sampled teachers was observed in the 1983/4 session. For the ROPS, in the 1984/5 session, the same teachers were asked to teach the topics in the syllabus, assuming that their students would answer the ROPS type of question in the school certificate examination.

Verbal interaction in the classroom was tape-recorded and transcribed. Although the study was concerned primarily with the discovery and correction of misconception, the units of interaction contributed by the pupils were calculated. This was because, although verbalization does not necessarily mean understanding, some educationalists (Barnes *et al.* 1969; English, 1981) insist that the conventional classroom which is dominated by teacher talk should be discouraged. They argue that increased pupil language is a step towards pupil understanding because it is only through verbalization that pupils' inner speech can be externalized, and then inconsistencies will be tackled. The units of thought in the lessons were segmented following the Aschner-Gallagher system (1965) for categorizing classroom verbal interaction. The discovery and correction of misconceptions were ascertained by eliciting evidence from teacher and pupil interaction. Three categories of this were sought, as follows:

(i) No hint of misconception.
(ii) Misconception discovered but not corrected.
(iii) Misconception discovered and corrected.

'No hint of misconception' was a situation of no indication at all that the concept taught was not grasped. 'Misconception discovered but not corrected' showed discovery of mistakes without a deliberate attempt to solve them. Thus when teachers detected mistakes made by students they either provided the right answers or elicited them from individual students or in chorus without giving any hint as to how the answers were derived. Correcting the misconception implied a deliberate attempt to make the learner understand his/her mistakes.

Types of misconceptions were also analysed. For the purpose of this study three types were sought: basic, habitual, consequential. Basic misconception was used to denote misunderstanding of basic ideas such as scale, gradient, drawing of cross-section, contour forms, as well as use of lay terms. Habitual misconceptions are lapses committed owing to failure to use a particular inevitable skill, for example measurement and the key. Consequential misconception differs from basic and habitual in that it is teacher-committed or the teacher accepts as accurate errors committed by students. This is a flaw committed by the teacher and the root cannot be tagged to mere lapse or ignorance. The frequency of lessons containing these types of misconceptions was calculated. Verbatim reports of the classroom interaction were made. The hypotheses were tested with t and Z tests of differences between correlated and uncorrelated proportions. Significance was tested beyond the 0.05 confidence level for a one-tailed test.

RESULTS

The results obtained from the analysis of the classroom verbal interaction are shown in tables 4.1 to 4.4.

Units of thought

The results on units of thought shown in table 4.1 were used in answering research question 1 and for testing hypothesis 1. The results in table 4.1 show that the proportion of students' units of thought in the ROPS (0.5331) is higher than in the WAEC (0.4248). The Z values for related proportions, which are $Z = 13.59$, showed teacher-dominated talk in the WAEC and ($Z = 13.21$) student-dominated in the ROPS. The Z test of differences between units of thoughts of students in the ROPS (0.5331) and the WAEC lessons (0.4248) for independent proportions was 14.51. This value is significant beyond the 0.05 confidence level. The hypothesis that students' talk will be higher in the ROPS than in the WAEC lessons is not rejected. This means that students verbalized more in ROPS lessons than in WAEC lessons.

Table 4.1 Proportions of units of thought in the lessons

Lesson type		Frequency	Proportion	Z test of proportions	Unit of thought per minute	Time (minutes)
WAEC	T	4,587	0.5752		4.24	
	S	3,388	0.4248	13.59*	3.13	1,082
	Total	7,975	1		7.37	1,082
ROPS	T	4,629	0.4669		4.06	
	S	5,286	0.5331	13.21*	4.63	1,141
	Total	9,915	1		8.69	1,141

T teacher, S student.
* Significant beyond the 0.05 level for a one-tailed test.

Discovery and correction of misconceptions

Table 4.2 shows the results obtained. These results answered research question 2 and were used for testing hypothesis 2. The data in table 4.2 show a higher proportion of lessons in which misconceptions discovered were corrected in the ROPS than in the WAEC. The reverse obtained for 'No hint' and for 'Misconceptions discovered and not corrected'. The t value (4.28) of 'Misconceptions discovered and corrected' in ROPS and WAEC teaching, which tested hypothesis 2, was significant. This shows that ROPS lessons encourage externalization and correction of misconceptions more than WAEC lessons.

Types of misconceptions

With regard to different types of misconceptions, tables 4.3 and 4.4 provided information for answering research question 3 and for testing hypothesis 3. Table 4.4 shows that a higher proportion of ROPS lessons exposed misconceptions than the WAEC. The proportion of misconception observation in ROPS lessons (table 4.4) is also significantly higher than in the WAEC.

Table 4.2 Discovery and correction of misconceptions

Frequency of lessons

Lesson type	No hint	Misconception discovered but not corrected	Misconception discovered and corrected	Total
WAEC	12 (0.4000)	17 (0.5667)	1 (0.0333)	30
ROPS	4 (0.1429)	9 (0.3214)	15 (0.5357)	28
t test	2.19*	1.88*	4.28*	58

( ) Proportions.
* Significant beyond the 0.05 (df = 56); a one-tailed test.

Table 4.3 Frequency of lessons with different types of misconception

Lesson type	Basic	Habitual	Consequential	Total
WAEC	19 (0.9500)	0	1 (0.0500)	20 (0.4000)
ROPS	23 (0.7667)	4	3 (0.2333)	30 (0.6000)
t test			1.73*	1.41

* Significant beyond the 0.05 confidence level for a one-tailed test.

The proportion of habitual and consequential misconceptions in ROPS lessons (tables 4.3 and 4.4) is also significantly higher than in the WAEC. Hypothesis 3 is therefore not rejected. This means that ROPS lessons reveal more misconceptions than the WAEC. Also ROPS lessons revealed other misconceptions (habitual and consequential) and WAEC lessons did not.

Table 4.4 Frequency of various types of misconception in the lessons

Element of misconception	Frequency		Proportions		Z test
	WAEC	ROPS	WAEC	ROPS	
Basic					
Scale	9	10			
Conversion	1	1			
Use of lay terms	1	14			
Calculation of gradient	1	3			
Difference between vertical interval, vertical distance and scale	–	2			
Identification of topographical features	33	16	0.9787	0.7465	
Meaning of contour spacing	–	3			
Identifying height of land	–	3			
Direction	1	1			
Habitual					
Use of key		15	0.0213	0.2535	4.16*
Consequential					
Lapse rate	1	1			
Misinterpretation of gradient ratio	1	1			
Misinterpretation of use of tunnel		1			
Total N = 118	47	71	0.3983	0.6017	2.21*

* Significant beyond the 0.05 confidence level for a one-tailed test.

DISCUSSION

Table 4.1 shows that students verbalize more during ROPS lessons than in the WAEC. The reason is that the expected question in the certificate examination, which is issue-based, induced teachers into providing opportunity for greater student talk through asking open-ended questions which needed expressive rather than transactional talk. Examples of exercises on the WAEC and ROPS by the same teacher on measurement of distance show clearly the differences.

WAEC. What is the distance along the major road from Brukino to Bosuso?

ROPS. Assume you live at Brukino and you are travelling to Bosuso within the shortest possible time. Which of the two routes will you take: the major or the secondary road? (Okpala, 1987:220-53)

While the question on the WAEC is theoretical the ROPS is realistic and requires decision-making. Making such a decision requires working, further questioning, and clarification of ideas. There is therefore greater involvement by pupils. WAEC lessons are mechanistic, as a teacher works out the solution to the question with the whole class and confirms understanding by posing the question 'Is it clear?' which is often followed by the response 'Yes'. In ROPS lessons students become involved participants of their own volition rather than by compulsion as they raise further questions on the issue. Educationists agree that this type of participation facilitates learning, as it stimulates critical thinking, is intrinsically motivating, and evokes a desire to seek relationships and discuss reasons.

Discovery and correction of misconceptions

Table 4.2 shows that the WAEC lessons conceal misconceptions as indicated by 'No hint' more than the ROPS as well as having a higher significant proportion of 'Misconception discovered not corrected'. On the other hand ROPS lessons had higher significant proportions of 'Misconceptions discovered and corrected'.

'No hint' of misconceptions which predominate in WAEC lessons could be interpreted to signify a state of ignorance which Vygotsky (1962) terms 'conceptual vacuum'. The teaching in WAEC lessons was dominated by teacher talk (table 4.1). This situation does not give opportunity for externalization of students' inner thoughts and for detecting misconceptions. An example of interaction in such a WAEC lesson is reported below.

Measurement of distance (revision)

Teacher. Assuming you want to measure an irregular course and you want to know the actual distance between A and B. Between A and B, assuming it is a river course. It could be a river course or a winding road. Then you make use of a piece of thread. Use a piece of thread, lay it over it. After that, what do you do?
First student. You measure.
Teacher. You measure to the scale, to the linear scale. That is, you have 0, 1, 2, 3, 4. These are in kilometres. Whatever is obtained, like it were a piece of thread, and you want to measure directly from there. You have this as your starting point. That means you stretch it all along that way (*demonstrating*). Okay. You are looking at me, anyway.
Students. No, your back is in the way.
Teacher. ... Okay, look at the board. What you are advised is to get the distance very well so that if you use a piece of thread all round the curves along the course later you transfer the distance to your scale and, okay, here fortunately I got four. Which means that on your scale below the topographical map the distance between A and B is four kilometres. So that if this were the scale - this is just for demonstration - that is how you measure irregular distance. You can use a piece of thread, a straight edge of paper. What can you use, again? You can use a straight edge of paper and a piece of thread. Yes. What again in measuring along irregular distance?
First student. We can use broom. We can use broom, broomstick.
Teacher. When you are measuring irregular courses?
Second student. Thread.
Teacher. Yes, what again?
Third student. Paper.
Teacher. How do you use that one?
Third student. By folding it.
Teacher. Do you fold it?
Third student. You do it like this (*demonstrates*).
Fourth student. You can make use of the edge.

Teacher. The edge, okay, how do you do it? Tell me exactly how you use it.

First student. Like this (*demonstrating, adjusting the sheet*).

Teacher. You transfer the curve on straight edge of a paper. Assuming you want to measure this curve. Okay, let's make it up to this end (*marking off the end*). So I can start and I mark (*demonstrating*). You see - okay, now you try to get just a bit of length at a time. So I put it there, then I fix it. Then I do it again and I twist it and twist it again. I twist it to follow the path that is straight, then I twist it again and again; I put, I twist, and then I put it down, I continue until I gain, till I get the measurement, I twist and I rotate and I twist. Okay, let me now get it up to this length, now I have transferred the curve on to the straight edge. Have you seen it? The next thing you do is to take it to the scale. Use the scale. If the linear scale is there (*refers to linear scale drawn on the chalkboard*) then place it at zero (*measures*) and then it gives you -?

Students. Two and a half.

Teacher. One, two, okay, let us assume it is 0.5, so that gives 2.5 kilometres, miles, feet, or whatever scale you use as linear scale. Are you all right with that?

Students. Yes. (Okpala, 1987:257-8)

With the positive response at the end of this interaction the teacher assumed that all students understood how to obtain the actual distances using the linear scale. There is no evidence of misconception. But did the students really understand? In such a class teaching situation it is most likely that some misconceptions are concealed but nobody likes to be the fool. Teacher discovering mistakes and supplying the right answers without actually making clear the conceptual error is a little improvement on 'No hint' but is useless without regard to facilitating conceptualization.

With regard to 'Misconceptions discovered and corrected', while the incidence is significantly higher for the ROPS than the WAEC, a major difference is that the former is pupil-orientated and the latter teacher-orientated.

In WAEC lessons student talk is mainly a response to teachers' questions. Corrections are made predominantly by the teacher or a bright student. The confused student is rarely involved. There is no room for further questioning or clarification. An example is show below.

Teacher. Show me a spur. Find as many as you can. I will go

round and you show them to me.
First student. That's it.
Teacher. From where to where?
First student. Here (*pointing*).
Teacher. Very good. Show me another spur.
First student. Here (*pointing*).
Teacher. Yes. You (*talking to another student*): where is a spur?
Second student. Here (*pointing*).
Teacher. No, no. That's a valley. You see that that's an inverted V shape. A spur has a V with the lower numbering at the bottom of the V. Now look at these two (*pointing at a spur and then a valley*). Now all of you look at the board. The best way to differentiate, to distinguish between a spur and a valley is taking note of the numbering. A spur has a V shape with the lower numbering at the apex. Valley on the other hand is an inverted V with the highest numbering at the apex (*illustrates*). It is clear?
Students. Yes. (Okpala, 1987:317)

In the ROPS lessons the mode of discovery and correction of misconceptions differ from the conventional pattern. In the free teaching-learning environment which existed in the ROPS lessons both teachers and students discovered misconceptions and demanded their clarification. A significant improvement on the conventional approach is the students' role in discovering and correcting misconceptions during ROPS lessons in group work as well as in class organization.

In class teaching during ROPS lessons the atmosphere was less tense than during WAEC lessons. Students of their own volition paired up and observed topographical maps, discussed the problem posed by the teacher, asked questions, detected misconceptions, and attempted correcting them. An example is reported below.

In teaching calculation of gradient, teachers often give the formula VI/HE. Some students make the mistake of regarding the VI as Vertical Interval rather than the difference between the highest and the lowest points. When the teacher discovered this misconception, he called the attention of the students and tried to correct the mistake.

Teacher. No. VI there means difference in height. So what you do to get this difference is to subtract the highest height from the lowest. Is that clear?
Student. Yes, ma.

The teacher then continued with teaching. Then the same student who was corrected went into an interaction with another student, questioning how the VI was obtained.

> *The same student.* How did you get 500?
> *Second student.* To get the height you've got to find the VI -vertical interval. What is the vertical interval, for a start?
> *First student.* 100 feet.
> *Second student.* You see, the land is rising.
> *First student.* Which way?
> *Second student.* This way (*pointing*).
> *First student.* Oh, that is, eh (*interruption by teacher*).
> (Okpala, 1987:303).

As the interaction took place in a class situation and was unauthorized by the teacher, though not prohibited, the two students were called to attention before they could work out the vertical difference. In group-work situations where student-student interaction was encouraged, the least misconception discovered was extensively questioned and clarified.

Questioning and demand for further clarification of ideas which occur in student-student interaction during ROPS group-work organization contributes immensely in externalizing further misconceptions. The flaws externalized include misinterpretation of features, arbitrary interpretation, and use of lay terms. In the case of misinterpretation, other students insisted on the interlocutor showing them the feature on the map. The problem of arbitrary interpretation is worse than misinterpretation because in such cases the interlocutor vocalizes what he/she knows without studying the map. For example, in interpreting the relationships between relief and settlement, students often generalized that thicker population exists in valleys and lowlands. In an instance (Okpala, 1987:335-7) in which population was concentrated on a pass, a student during a group discussion made the usual generalization. The members of the group disagreed with the contributor and insisted during a two-minute discussion that he should identify the valley by asking him questions such as 'How do you know that a place is a valley?' 'Show me what a valley is. Explain it.' 'Show me a valley.' 'Now, is this a valley?' (pointing).

Feedback to the questions by the interlocutor showed attempts at concealing a conceptual vacuum through rigmarolling, regurgitation of facts, and the use of lay terms, e.g. 'Look at the contour lines, look at the contour lines. Now, okay, look at ...'' 'About those valleys. Look at those somethings. Right, okay, just look at. Just see. I won't say on the eastern side of the map we agree that they are sort of situated in valleys, right.' 'A valley is a lowland between two highlands. Right ...' In the end the interlocutor was pinned down to closer observation and he started reconstructing his ideas. He observed, 'There is a highland' (pointing). 'Okay, look at' (pointing) 'this is a valley. Okay, see the height, 1,250 ft. But look at the height: 650,600.'

Students use lay terms such as 'the area is normal', 'the area is effective', 'the something slopes downwards', 'the land is not good', 'the place is a very good something', 'the land is not stable', 'everything', 'there', 'here', 'in that place'. These terms were not often used where students lack words to explain the topographical feature observed. Verbalization in ROPS (table 4.4) externalized these lay terms. Through teacher-pupil and pupil-pupil probing students were led into discovering the appropriate terms.

Types of misconception

Table 4.4 shows that ROPS lessons elicited the three types of misconception - Basic, Habitual, and Consequential - while the WAEC elicited Basic and Consequential. The WAEC did not externalize evidence of lack of use of the key. Use of the key is important for effective interpretation of topographical maps. Quite often students make arbitrary offhand interpretation without reference to the key. These were exposed in ROPS lessons because knowledge was put to use and specificity was required.

In ROPS lessons the problem given below was presented to students in three groups (A, B, and C) to solve:

Topic: measurement of distance

The Problem
It is said that the shortest distance between two towns is the straight-line distance between the two towns. Do you think that the class 1 road that runs from Bosusu to Begoro passed through the best path? Give reasons for your answer.

Students worked in groups and at the end of ten minutes presented their solution to the other groups. Group C in presenting their work read thus:

> *Group C leader.* And besides in cases of, if the road is straight, from Bosusu to Begoro in cases of emergencies it will not be easy to transport the accident victims down to the nearby resting places or hospital. So the class one road down from Bosusu to Begoro leads the best way.
> *Teacher.* Any questions to ask group C?
> *Group A member.* Let me ask one question. I have been looking around here for a hospital. There is no hospital. Where is the hospital sign? There is no hospital.
> *All group C.* There is.
> *All students (start searching through, referring to the key.)*
> *Group A member.* Okay, where is it?
> *Group B member.* Where is it?
> *Group A member.* There is no hospital there. It is a guest house. Rest, rest house. So definitely that statement is wrong.
> *Group C member (searching through the map).* No, there is at this *(stops and starts searching).* There is a point of correction.
> *Group C members (put heads together and talk in low tones).*
> *Group C leader.* I mentioned something like cases of accidents or emergency cases that it would be easier to find rest house or hospitals. But if you take a straight path from Bosuso to Begoro the rest house there is down at Begoro and if you take the class one road if you look carefully there is no rest house. So definitely if the path goes straight as far as for cases of accidents and emergencies there will be only a rest house at Begoro, so for that point the straight road will be good.
> *Teacher.* Any question on that?
> *Group B member.* What is a rest house? Just for relaxing. So even in the case of accidents where can you report?
> *(The class are silent).*

The interaction reported above shows that ROPS teaching exposes and disentangles misconceptions which are sometimes complex in nature. The example above led to close observation of the key as well as exposing group C students' ignorance of the meaning of 'rest house'.

Boardman and Tower (1980) recognized that ability to read symbols is one of the basic skills in map reading. They reported that the major cause of inadequate use of map symbols among school certificate students was that, when this skill is introduced, as it often is, at the beginning of the school certificate course, then there is a tendency for the skills to be neglected in subsequent years. They stressed the need for the

Talking and problem-solving

skill to be repeatedly practised and reinforced in subsequent years.

Although basic misconceptions were externalized in both the ROPS and WAEC lessons, some fundamental confusions were observed only in ROPS lessons. These were the differences between VI (Vertical Interval) and VD (Vertical Difference); and the meaning of contour spacing. The interaction on the problem of VI and VD is reported below.

Students were working in twos on a problem and a pair felt that calculating gradient was necessary. They first obtained the horizontal distance and were calculating the vertical difference when an argument broke out, thus:

First student. You have to subtract this (2,000) from this (4,300).
Second student. It is not true. The VI is 50 ft.
First student. I do not agree with you. What we need is the difference between the highest point (*pointing*) and the lowest point.
Second student. It is not right: in the textbook the formula is VI over HE. This is 50 upon.
First student. Which textbook?
Second student. Nimako.
First student (searches her bag anxiously). Oh, mine is not here. (*Calls out to another student*). Can I have your Ni?
(*The teacher intervenes.*)
Teacher. Ifeoma, is there any problem?
First student. Yes, ma.
(*Teacher moves over to the two students.*)
Second student. Ifeoma is insisting that the Vertical Interval is 50 ft. I was telling her that to get the VI you subtract the highest - I mean the lowest - distance from the highest one.
Teacher. You see, that's why I told you to use VD, that is, Vertical Difference. The difference in height of the distance whose gradient you are calculating. Where is the distance you are working on?
Second student. The highest point is 4,300 ft and the lowest is 2,000.
First student. 2,000 ft.
Teacher. What is the difference?
First student. The Vertical Interval difference is 4,300 - 2,000. That is 2,300.
Teacher. 2,300 what?
First and second students. Feet.

Reference to the basic text (Nimako, 1982:48) shows that the formula for gradient is VI/HE. VI is defined as the difference in height between the two places. In the same book (p.28) it was stated that 'The vertical interval between one contour line and the next is called the vertical interval (VI)'. It is therefore most likely that students battle over this dual meaning in the WAEC lessons but that owing to limited opportunity for verbalization the misconception is concealed. In calculating gradient, the problem of scale which involves calculation of distance on the ground is overt, and it is most likely that students who could not calculate gradient would encounter difficulty in finding the vertical difference. It was only in ROPS lessons that this problem/confusion in finding vertical difference was made explicit.

In teaching contour forms in the conventional classroom, teachers generally started by giving the definition of contour, after which they gave students examples of sketches of topographical features. In two ROPS lessons teachers attempted teaching with models. After contour forms had been inserted on the models, students were required to map out the contoured models. Problems arose when they were required to show steeper slopes with closer contour lines.

Students questioned consistently the reason the contour line should be made closer on the steeper side even though the height of the land was the same (Okpala, 1987:231, 362). Although in both instances the teachers failed to correct the misconception, as they emphasized the dictum 'the closer the contour lines the steeper the slope', the discovery of this misconception reveals the root of some students' confusion and inability to interpret topographical maps.

The higher incidence of consequential misconception observed in ROPS teaching supports the efficacy of ROPS teaching as an avenue for providing opportunities for improving learning. Although these flaws committed and accepted by the teacher were not questioned in the quasi-experimental situation, it could be argued that if ROPS questions are adopted in WAEC map work and pupils get accustomed to probing, in the near future such flaws will not escape pupil enquiry.

EDUCATIONAL IMPLICATION

A major educational implication which emerges from this study is the evidence that setting Reality Orientated Problem Solving questions in final examinations could influence teachers to reconstruct classroom language. Expectation of ROPS questions induce teachers into involving goal-directed learning and creating a freer atmosphere which results in greater pupil talk and pupil-pupil interaction. These facilitate exposing misconceptions in a form that would facilitate meaningful correction and learning.

The discovery of misconceptions, other than basic, particularly the lack of use of the key, is significant in the teaching of map work in schools. It reveals to teachers problems which contribute to students' poor performance in the subject.

With regard to aspects of basic misconception, findings from this study draw teachers' attention to the root of misconceptions in understanding topographical features. This is a lack of understanding of the meaning of the terms. The most significant among this is the meaning of contour lines. As the contour line is the seedbed of topographical information, it could be argued that where it is misconceived the reading and interpretation of topographical maps are reduced to parrot learning. The same argument should be extended to problems in calculation of gradient and scale. The implication of an increase in the incidence of consequential misconception for the teacher is sharpening his/her awareness of the challenges in teaching for ROPS questions. The teacher should be able to cope with incidental issues and misconceptions which occur in the course of the lesson.

RECOMMENDATIONS

1. There is a need to set realistic questions in map work in school certificate examinations as a means of inducing productive and meaningful classroom interaction which facilitates externalizing pupils' misconceptions in map work.

2. Teachers should assist students to cultivate the habit of making reference to the key in all geography classes. The practice of teaching use of the key in the initial school certificate class and not paying attention to it in subsequent classes should be reorientated.

3. There is a need for geography educators to derive practical ways of leading students into conceptualizing the

meaning of contour lines, scale, and gradient. It is likely that many problems which students encounter in interpreting topographical maps are caused by lack of understanding of these concepts.

4. Group work is highly recommended as a means not only of enhancing externalization and correction of misconceptions but also of enhancing pupil-pupil learning.

FINAL WORD

The findings from this study support setting Reality Orientated Problem Solving types of questions in school certificate examinations as a means of producing goal-directed verbalization which can enhance discovering and correcting misconceptions. It is in line with the idea that expressive talk encourages verbalization of inner thought, and this facilitates clarification of ideas so that students come to realize what they do not understand. Although the result of this study is in favour of ROPS questions as a means of inducing productive verbalization in map work, it should be realized that, unless the examining bodies incorporate the ROPS pattern of questioning in school certificate map-work examinations, improving verbalization and conceptualization in map work will remain a dream.

REFERENCES

Aschner, M.J., and Gallagher, J.J. (1965) 'Aschner-Gallagher system for classifying thought processes in the content of classroom verbal interaction', in S. Anita and G.E. Boyer (eds.), *Mirrors for Behaviour: an Anthology of Classroom Observation Instruments*, Michigan: Research for Better Schools Inc.

Barnes, D., Britton, J., and Rosen, H. (1969) *Language, the Learner and the School*, Harmondsworth: Penguin.

Boardman, D. (1983) *Graphicacy and Geography Teaching*, London: Croom Helm.

Boardman, D., and Towner, E. (1980) 'Problems of correlating air photographs with Ordnance Survey maps', *Teaching Geography* 6,2:76-9.

English, M. (1981) 'Talking: does it help?' in C. Sutton (ed.), *Communication in the Classroom: a Guide for Subject Teachers on the more Effective use of Reading, Writing and Talk*, London: Hodder & Stoughton.

Ghaye, A. (1986) 'Outer appearances with inner experiences: towards a more holistic view of group work', *Education Review* 38, 1: 45-56.
Lunnon, A.J. (1969) 'The understanding of certain geographical concepts by primary school children', unpublished M.Ed. dissertation, University of Birmingham.
Macintosh, H.G. (1970) 'A constructive role for examination boards in curriculum development', *Journal of Curriculum Studies* 2: 32-9.
Marsden, W.E. (1976) *Evaluating the Geography Curriculum*, New York: Oliver & Boyd.
Okpala, J.I.N. (1987) 'The feasibility of Reality Orientated Problem Solving questions in WAEC examinations as a means to improving the teaching and learning of mapwork in Nigerian secondary schools', unpublished Ph.D. thesis, University of London, Institute of Education.
Owen, J.G. (1973) *The Management of Curriculum Development*, Cambridge: Cambridge University Press.
Piaget, J. (1959) *The Language and Thought of the Child*, London: Routledge & Kegan Paul.
Slater, F. (1979) 'The role of language in the geography lesson', *New Zealand Journal of Geography* 67: 18-19.
Slater, F., and Spicer, B. (1980) 'Language and learning in a geography context', *Geographical Education* 3: 477-87.
Slater, F. (1982) *Learning through Geography*, London: Heinemann.
Vygotsky, L.S. (1962) *Language and Thought*, Cambridge, Mass.: MIT Press.
West African Examinations Council (1975-84) *Chief Examiners and Moderators Reports*, Lagos: West African Examinations Council.
(1982) *Questions on Mapwork in WAEC Examination Paper 1B*, Lagos: West African Examinations Council.

Chapter Five

LANGUAGE AND LEARNING IN MULTICULTURAL EDUCATION

Daniel Lewis

PERSPECTIVES ON LANGUAGE IN MULTICULTURAL EDUCATION

I use the term 'multicultural education' as an inclusive one involving all issues of race and culture, including approaches that emphasize respect for cultural diversity and those that stress the need to combat racism. Within this broad concern are three complementary perspectives about language: cultural pluralism, equality of opportunity, and anti-racism.

Cultural pluralism emphasizes an understanding and respect for cultural diversity. This should mean that the home language is valued by the school, whether it is a separate language such as Urdu or a dialect such as Creole-influenced English. In some schools bilingual policies have been introduced and Creole is used in selected situations. This can help children develop a positive attitude towards school and give a good foundation for the learning of standard English.

Equality of opportunity. Children from ethnic minorities need to be fluent in the different styles of standard English, especially writing, so as to give them an equal chance of succeeding in the education system. Within multicultural education, the main strategy is to recognize the basic linguistic competence of ethnic minority children, and develop it towards standard English, through participation in the mainstream classroom. There is a shift from seeing the pupils as a 'problem' and 'deficient' towards a view that schools need to change to meet their needs.

Anti-racism aims at combating prejudice and discrimination, relating them to the structure of power in society. If the children of ethnic minorities can become fluent in standard English, the language of power in our society (Searle, 1983), this can help them change the system. Also, by developing a critical view of the use of language in society, prejudice and discrimination within school and society can be challenged. An effective anti-racist language policy is important for the motivation of ethnic minority children and for the general education of all pupils.

THE GEOGRAPHY CURRICULUM

All three perspectives in multicultural education are needed for an effective policy. For geography teachers they provide the background of two major concerns:

1 *Multicultural classrooms.* There are special language needs of students from ethnic minorities. The curriculum should validate the home language of pupils and, at the same time, ensure they achieve competence in standard English.
2 *Prejudice.* There is the role of language in the process of challenging and changing racist attitudes and behaviour. This involves an understanding of the values implicit within language and the use of language activities in the classroom to criticize racism and encourage change.

These concerns will be looked at under four headings: (1) dialect, bilingualism, and the development of standard English, (2) general language strategies: establishing a climate of hospitality, (3) specific language strategies: open and structured approaches, (4) language to challenge racism.

DIALECT, BILINGUALISM, AND THE DEVELOPMENT OF STANDARD ENGLISH

Teacher attitudes

Edwards (1983) argues that the attitude of teachers is the critical factor in the development of language in multicultural classrooms. Underlying these attitudes may be an ignorance about language, and hidden values. In this section I shall briefly outline the main points that geography teachers need to know as a background to more specific strategies.

Standard English

Standard English has no linguistic superiority over other dialects. Traditionally, standard English was seen as a superior form of English, and other dialects were devalued. In fact standard English is itself only a dialect and has no linguistic superiority in its grammar, vocabulary, or pronunciation, compared with other dialects (Trudgill, 1975). Although it is socially more important, this does not mean that the home language of the child should be negatively judged.

Standard English is sometimes too narrowly defined. Children have to know the grammar of English and its appropriate use but there are many aspects of the language which are open to discussion. Crystal (1984) points out that English is a living language which changes as society does. There are debates over traditional rules and the appropriateness of the style used. Sometimes geographers insist on using a scientific genre in writing (Kress, 1982) but there may be other styles that could be used, depending on the task set. Different styles of standard English will be appropriate in different situations.

Creole

'West Indian Creoles are perfectly regular, rule-governed linguistic systems' (Edwards, 1983). Creole has its vocabulary mainly from English, but the grammar derives from West Africa. It is not bad English but a dialect, furthest removed from standard English, with its own internal validity. Some experts would actually classify it as a separate language. It is important that Creole-influenced pupils do not have their home language devalued as deficient English.

Speaking a Creole-influenced dialect need not hinder the development of standard English. 'Interference' is a term used to describe how the first language influences the learning of a second one, and is familiar to most people who have tried to learn another language. It is possible for Creole grammar to 'interfere', but this does not seem too significant in the classroom. Most Creole-influenced pupils were born in Britain and can shift the way they speak for different situations, such as home, school, and talking to friends. Also, Richmond's (1979) analysis of their writing suggests that most 'errors' are similar to those of other children, with dialect having only a small influence.

Bilingualism

Bilingualism is an asset rather than a liability. In many parts of the world it is normal to speak two or more languages fluently. Bilingual children born in Britain have the potential to grow up speaking the mother tongue and English without being held back in either language. 'Interference' is more likely where the pupil is a recent arrival to the country (Jones and George, 1981). Nevertheless, Dulay *et al.* (1981) argue that interference is not as significant as was once thought, and the difficulties faced by second-language learners will be similar to those of other children. The implication is that being bilingual need not be a problem. It is possible to develop expertise in both languages without their hindering each other.

Errors by pupils may be positive indicators that they are making progress in their language development. As children move towards adult standards they pass through a sequence of stages. At each new stage they experiment with the language, making 'errors' in the process. These 'errors', then, may indicate where the child is trying to improve. If the teacher marks them wrong, it may be counter-productive, because it discourages language exploration and fails to give credit for progress made. The main difference between first and second-language learners will be due to the lack of experience in English of some bilinguals, especially recent arrivals. This will put them at an earlier stage of language development and they may need extra help. Sometimes these children may have made progress in their conceptual development but find their language skills hold them back.

The special language needs of ethnic minorities

From the foregoing we can see that many of the language needs of bilingual and Creole-influenced pupils will be similar to other pupils'. The main difference is that some of these students may have a longer journey to go on, moving from the language and culture of home to the language and culture of school. The cultural and language shift from home to school may be greater than for other pupils. Improving the language of these students cannot be achieved by concentrating on separate language skills in isolation. Language and culture are closely connected, so that progress is only likely where there is a commitment to multicultural and anti-racist education. The attitude of the teacher is critically important. There has to be a classroom ethos that encourages respect for the home language and accommodates the different stages that pupils are in, as they move towards competence in standard English.

GENERAL LANGUAGE STRATEGIES: ESTABLISHING A CLIMATE OF HOSPITALITY

An effective language policy requires a classroom ethos that respects the existing language of the children, encourages exploration and experimentation, and accommodates their diverse language needs. Levine (1982) calls this a climate of hospitality. Three general strategies for achieving it are (1) validating the home language of the children, (2) using a marking policy that gives children the confidence to explore language, (3) developing 'mixed ability' teaching that can cater for the varied language needs of a class.

Validating the home language

A familiar theme in geography teaching is to show the relationship between everyday life and the world. Food, consumer products, and other items are found to come from various countries, which can be located on a world map. Language can also illustrate international links. In one class I visited the children did a class survey of languages spoken by pupils and then related these to a world map on languages. Another example is the *World Studies 8-13* Teacher's Handbook (Fisher and Hicks, 1985), where there is a class activity called the Word House. Children work in pairs, collecting words from each other from different word 'families' that have contributed to modern English. Also, in many schools today, courses are being established in 'language awareness', where geographers could make a contribution. Two interesting books are Raleigh (1981) and Houlton (1985). Geography teachers could also validate the home language by having signs and displays in different languages and by using bilingual materials such as the ILEA (1981) packs on London.

Developing a marking policy

A knowledge of general language development (Dale, 1976) helps to ensure that errors can be marked in a positive way. Table 2.2, for instance, gives a simple description of earlier and later stages of writing that I have found useful. This approach takes time to develop, and the teacher may become hesitant about making any language corrections at all. This applies not only to writing but also to talking and reading aloud. Interrupting and giving the correct pronunciation or word may inhibit the child from trying another time. The action taken by the teacher depends on the student, the situation, and the task involved, and needs to be part of a general policy on marking and correction. Dunsbee and Ford (1980) suggest that teachers should concentrate on the meaning first, and then look at the adequacy of conveying that meaning through selective marking. One answer for geography teachers is to mark for meaning only and do no language marking at all. In the short run this may be the best policy for a teacher who is unsure about language development. It is certainly better than the negative effect of insensitive correction or meaningless advice such as 'Write in proper sentences', or 'Improve your spelling', which give no practical help. In the long run, though, as more is found out about language, a selective and positive marking system can be developed.

Allowing for different levels of language development

Within any classroom children will vary in their abilities and achievements, including their language skills. The term 'mixed ability' is often used to describe this variety and is most marked in comprehensive schools that have abolished streaming, setting, and banding. Teachers have to develop strategies to cater for the range of abilities, including children at different stages of language development. How can this range be catered for? There are two main ways of achieving this: the first is an _open_ approach which aims to include all the children in similar open language activities; the second is a _structured_ approach which emphasizes different resources and activities for different children. Teachers often combine both approaches in the classroom.

The open approach

Here all the students in the class are involved in similar kinds of language activity. Different levels of development can be accommodated through the openness of the activities.

1 Practical activities involving a lot of discussion and collaboration.

2 Open tasks, especially those that can involve more personal experience and imagination.
3 Resources that can be used by children in different ways at different levels.

The open approach aims to avoid the problems of streaming within the classroom, especially where this isolates bilingual children in 'remedial' groups. Children from ethnic minorities have the potential to be successful achievers but are held back by their language skills. Giving these children remedial or easier tasks may lower expectations and fail to give them the experience to improve. In contrast, the open approach will encourage the use of 'mixed ability' groups so that children with language difficulties are distributed evenly round the class. These groups can be used for discussion and collaboration between pupils, with all the children participating. Good examples of these kinds of activities are found in the Collaborative Learning Project (Scott, 1985).

Another idea is to give children open writing tasks that allow them to write varying amounts appropriate to their level. This is easier where a range of writing is acceptable, as in personal and imaginative writing. A comparison can be made with the 'neutral' question for GCSE examinations which allows a variable response from the candidate. A problem is where pupils do not take advantage of this openness, holding back from discussion or extending writing. Pupils used to writing short, brief answers may continue doing this, despite the opportunity to write more. A third way is to use resources that all children can benefit from. Books, using many photographs and pictures, and television programmes are particularly valuable. Designing or finding these kinds of resources, though, is difficult.

The structured approach

This is where work is structured at different levels of difficulty or according to need. Students either do the same work, with higher achievers moving quickly through easier tasks, or different tasks are set for different children.

1 Separate tasks allocated to groups and individuals, according to level of achievement or need.
2 Stepped questions of increasing difficulty.

The structured approach involves children in different

language activities according to their level of development. It aims to overcome the problem of giving the children the same work, which may be too difficult for some and too easy for others. A traditional way of doing this is to set different levels of work for different children. This can involve different resources and different tasks. The easier questions involve resources at the lower reading level, probably with more pictures, and considerable help with the vocabulary and language structures needed. This grading of work can be individualized or linked to groups in class. Examples are given in the next section on specific language strategies. The main criticism is that this is a form of setting and will lead to underachievement by the least successful groups because of low expectations and low esteem. These groups may include a high proportion of black children because of their complex language needs. Such a system could be a form of institutional racism, despite the best intentions.

An alternative method is to allocate work according to need and encourage a classroom ethos that respects diversity, so that all children feel included, regardless of their level of achievement. It is a principle of multicultural and anti-racist education that you treat children equally in value but differently in kind, according to need. The variety of needs will relate not just to ability but also to physical, emotional, language, and social needs, among others. The children from ethnic minorities will have particular needs that must be accommodated, including potential high achievers who may be held back because of language needs and lack of a supportive multicultural curriculum. In multicultural classrooms the term 'mixed needs' is more appropriate than 'mixed ability'.

If there is respect for cultural diversity in the classroom and it is normal to cater for the diversity of children's needs, then giving children different activities is acceptable. This may include work at different levels of difficulty. It is also a good idea to give the pupils opportunities to make their own choice about which work they do. Teachers may fear that the children will always choose easier or familiar work that does not stretch them, but this need not happen if there is a healthy classroom atmosphere. In a classroom that caters for diversity of needs, there will be many different language activities going on, and the pupils will have a say in what they do.

Another method is to set stepped questions of increasing difficulty. All the children can do the easier work at the beginning, which covers the main concepts and skills. This earlier work could be designed to include a range of language activities, so that the low achievers are not restricted in their development. If children are allowed to work at their own pace, this gives the slower workers the time to be successful in what they do and allows the higher achievers to be stretched. Once children have completed the core work they can go on to the more difficult work. This could include a range of tasks, with the children choosing what they want to do, or being guided to language activities where they need more practice. The easier language tasks make fewer demands on the children, either by giving them the vocabulary and language structures they need, or by using open questions that do not require precise answers on language skills. The reading materials can be graded, with the easier texts using a narrower vocabulary, simpler sentence structure, and less formal style, and avoiding abstraction and sophisticated analysis. If 'mixed ability' groups are used, this could correct the tendency of this approach to drift into streamed groups.

Both the open and the structured approaches require adequate time to allow the children to develop their language skills. The biggest pressure against this is the desire to cover the curriculum in a certain way, within a narrow time limit. The temptation is for the teacher to give all the children the same work to do, in a fixed time, with relatively closed comprehension questions. Differences in language development are not catered for, except in the slight openness of some of the questions and where some conversation is allowed, usually casual conversation. Some of the higher achievers may be given more varied writing tasks, if they finish the main work, but the general aim is to move rapidly through the topics on the syllabus.

Shifting towards teaching strategies that can cater for language diversity may require a shift in value about the nature of geographical education. The major problem is that teachers want to control what happens, so they do not allow the pupils to find out for themselves. Ideally a scientific approach should encourage exploration and experimentation; a humanistic approach should encourage personal expression and imagination; and a 'radical' approach should encourage discussion and debate. These are unlikely to occur if the teacher adopts a strong 'transmission' model of education (Barnes, 1975). This contrasts with an 'interpretation' model which emphasizes openness and uncertainty, and values the existing knowledge of the child.

Multicultural education

Case studies

The two case studies below illustrate ways of managing the range of language needs in a class. Both topics take on issues that are important in multicultural education and are relevant to geographical education. Both use open and structured approaches.

South Africa

The Second Language Learners in the Mainstream Classroom Project (Riley, 1986) has produced materials for use with the story *Journey to Jo'burg*, by Beverley Naidoo (1985). The aim is to understand the Apartheid system of South Africa. Although the project has developed this work for English teaching, the underlying principles are useful for geography lessons. Generally, an open approach is used most of the time, with some structured elements:

Open approach: all children can participate
1   The class is divided into 'mixed ability' groups.
2   Discussion is encouraged throughout the topic.
3   The work is introduced with a whole-class simulation on Apartheid.
4   The book is read together as a class and is accessible to most of the children.
5   Writing activities are open, including empathetic writing, dialogue, poems, opinions, and criticism.

Structured/open activities: combining approaches
In group discussion the pupils do preparatory work for the writing. They use typical techniques found in the teaching of English as a Second Language (ESL), as described in the next section. One method is to have sets of picture cards which have to be sorted into correct groups. The pictures can be used by bilingual pupils to help them in their writing. What is interesting is that these techniques are used mainly as a *preparation* for the open writing tasks. This contrasts with traditional ESL approaches, which often put more constraints on writing.

Structured approach: mixed-needs teaching
1   Audio-tapes are available for listening to the story.

2 The pupils often have a choice of activities.
3 It is also possible to get hold of summaries of the book in other languages, although, as far as I know, this is not part of the project.

Native Americans
The following work has been used with first-year secondary pupils. After doing a topic on the Plains, a study was made of Native Americans today, to challenge stereotypes, often based on westerns, and consider some contemporary issues. One of these is the demand being made of Native American land by mining companies.

Open approach: all students participate
1 Tape-slide on *Unlearning American Indian Stereotypes* (CIBC, 1977), followed by discussion.
2 The book *Apache Family* (Fear and Fear, 1985) is read to the whole class.
3 *Red Ribbons for Emma* (Preusch et al. 1981), which tells the story of a Native American's fight against the mining communities, is read to the class and discussed in groups.

Structured approach: core work followed by supplementary tasks
1 The first work sheet uses simple ESL techniques to cover the main concepts and issues. All the children do this work.
2 Some bilinguals still find the work difficult so a complementary work sheet is used to cover the same key ideas.
3 The second work sheet explores the key ideas and issues in greater depth using a variety of language activities. The teacher can direct pupils what to do, or give them some choice. As the basic information was covered in the first work sheet, this allows more open work to be explored.

SPECIFIC LANGUAGE STRATEGIES: OPEN AND STRUCTURED APPROACHES

The open approach of Language across the Curriculum

Although the structured approach has a role to play in the ideas of Language across the Curriculum, its main philosophy involves an open approach. The central idea is that language development comes about through the involvement of pupils in *meaningful* activities. Through discussion, interesting books, and opportunities to write in their own words the students are given an open situation within which they can explore language use. As all children have an innate desire to become skilled language users so as to communicate with others, the main task is to set activities that encourage their active participation. This will be helped by having *real* purposes for the language, not just communication with a teacher-as-judge.

From this point of view the structured approach may be inappropriate, as it fails to see language use as a holistic process. Practising skills, as in a spelling test, may not actually help children improve their spelling, even where they get good results in the test. Giving children books at the appropriate reading age may prevent them from extending their reading skills, especially if the books are poorly written and lack interest. In contrast, if you give children a range of books, they can choose which ones to read and, if they need to stretch themselves, an interest in the book will motivate them to try. For each topic the teacher should have a range of books that the children look at, browse through, or read.

Writing is unlikely to develop if pupils are given all or most of the words they need. They need the opportunity to write in their own words, at length, and sometimes to see their first go as a draft, needing revision and development. Always having closed questions with specific correct answers is likely to inhibit writing. Open questions give the students a chance to make their own contribution. This can be achieved in different kinds of writing, such as description, analysis, speculation, opinion, argument, stories, and personal writing.

Both reading and writing are helped by discussion. Classroom activities can be designed to encourage collaboration between students. The talking helps learning and can help reading and writing development through discussion of a text or ideas for writing. Generally the ideas of Language across the Curriculum will be applicable to multicultural classrooms. A good introduction to this is Sutton (1981). The following suggestions are a selection that seem particularly appropriate for ethnic minority children.

Talking and listening

New arrivals may need periods of mainly listening at first.

Planned group activities give bilingual students the opportunity to participate, e.g. the Collaborative Learning

Project (Scott, 1985).

Casual groupings, where conversation is allowed, give students experience in language use.

Students need opportunities to talk about their experience, opinions, and ideas.

One-to-one talking and listening ensure that all students participate.

Bilingual pupils should be encouraged to ask questions when they do not understand work.

Reading

There needs to be a range of reading materials on each topic.

These could include books thought suitable for younger pupils.

For each topic, teachers should organize a collection of books and give the pupils time to read them.

Lessons can be organized to encourage *active* reading using group discussion.

Writing

Allowing time and discussion before the writing is started.

Encouraging a range of writing activities (Slater, 1982; Lewis, 1987).

Having more open tasks, with opportunities for writing from pupils' own experience.

At present the National Writing Project (Richmond, 1986) is exploring many interesting strategies.

Structured language activities

A different method is to give the pupils more guidance in their language activities, which are broken down into specific skills. Some children, for instance, do not pick up the use of the full stop at the end of a sentence. They may spread full stops liberally around the page, hoping to get some in the right place. Another example is where a student may not understand the differences between description, analysis, or opinion without some help. Lunzer and Gardner (1984) outline a structured approach to reading. Using group discussion, the students are encouraged to analyse the text through carefully designed activities. One kind of task involves sequencing a set of paragraphs that are in the wrong order. Another kind involves finding the missing words in a paragraph. These can be worked out by concentrating on the meaning and logical progression of the text.

This structured approach is often used in special needs teaching and in the learning of another language. It can also be used in ordinary class work, but its overuse may restrict language development. It needs to be integrated or balanced with open approaches, as described in the case studies on South Africa and Native Americans. Lunzer and Gardner balance their activities, to some extent, by using open discussion as well as structural analysis. In my opinion the more open holistic philosophy of Language across the Curriculum should be the main approach, which can then be modified by more structured activities, where this is relevant to the objectives of the lesson and the needs of the pupils.

Structured language activities developed within ESL

ESL teaching sometimes uses highly structured exercises where the choices made by the students are limited to selected language activities. In making these choices the aim is to help the student think about the topic and learn a language skill at the same time. As the trend is towards the integration of ESL pupils into the mainstream classroom, increasingly geography teachers will have some responsibility for their education, and will find it necessary to know the main ESL teaching approaches. In my experience, teachers often find these methods useful, not least because they can be used to make sure the pupils have the correct information written down in their exercise books. The danger is that more open tasks are not developed. The following outline gives the main methods used.

Some of the techniques are:

1 Giving the students an *example or model* or a suitable sentence.
2 Giving the students *all the words they need* but getting them to put them in the correct order, e.g. the start and finish of a sentence or a sequence of sentences.
3 Giving the students the basic structure and *getting them* to work out some of the words, e.g. filling in the missing word, completing a sentence, changing sentences from one form to another, such as a false statement to a true one.
4 Gradually these can be made more complicated, e.g.

writing a paragraph describing, say, a photograph, using a certain structure and some given words.

These are described in more detail below. Some of the examples are based on the case study of Native Americans described in the previous section.

Students are given the words, phrases or sentences to select or rearrange

1. *Choosing the correct ending to a sentence*
   The Great Plains were once (a) forest, (b) desert, (c) grassland.
2. *Linking the start and finish of a sentence, e.g. term and definition: putting the correct definition with the term*
   Herbivore   eats   animals
   Carnivore   eats   plants and animals
   Omnivore   eats   plants
   Other examples are: Question and answer, cause and effect.
3. *Substitution table: linking the words and phrases in the columns to make proper sentences*

The Plains	were	on reservations
Native Americans	have	on part of the Plains today
Mining companies	grow	grassland, in the past
Crops, like wheat	live	low rainfall
	want	the Native American lands

4. *Ordering a sequence: putting sentences into the correct order*
   On the other side of the mountain the air descends and warms.
   The moist air rises and cools which causes rain.
   The warm, descending air is dry, so there is little rain.
   Moist air comes from the sea and moves towards the mountains.
5. *Classifying words into groups: put the words under the correct heading*
   Physical features   Climate   Vegetation
   rainfall, oak, fir, warm, wind, berries, rocky, ash, cacti, mountains, desert, dry, plains, rivers, etc.

6 *Choosing the correct labels or descriptions of pictures or photographs*
A Long hut made of wood.
B Pointed tent stretched over poles.
C Round hut made of wood and earth.
7 *Choosing whether statements are true or false*
The Great Plains were once grassland.
Buffalo used to eat small animals.
Early Native Americans had no horses.
The buffalo became extinct.

Students are given the basic structure but have to work out some of the words or phrases

A model sentence may be given as an example, followed by some sentences or paragraphs with words missing.

1 *Finishing a sentence or filling in the missing word - one-word response.* This construction is familiar to teachers. It is also possible to ask the student to work out more than one word in a sentence or paragraph. Three kinds of word can be missed out: specialist words; words that are important for the meaning of the sentence (Lunzer and Gardner, 1984); and words in a cloze exercise where there is more emphasis on language structure by deleting words at regular intervals.
2 *Finishing a sentence - more open response.* One example is just to give the first words of a sentence.
3 *Re-writing sentences*
   (a) Changing a false statement to a true one.
   (b) Changing the sentence from the first to the third person (or vice versa).
4 *Giving a structure for different kinds of sentence.* There are various kinds of sentence that you may want to help them to write. Examples: factual statements, causal links, statements of probability, description, comparison. The children are guided by giving them the structure and vocabulary, e.g. probability: 'I think this place may be located in ____.'
5 *Paragraph writing.* These methods can then be developed into paragraph writing, as shown in the following example.
   1 Give the pupils three photographs and a set of words describing them.
   2 The pupils work in groups or pairs, sorting out the

words appropriate for each photograph.
3. An outline paragraph is given out which pupils have to complete for each photograph: 'In this photograph I can see __ and __. The land is __ and the plants are mainly __. I think the climate is __ because __. The main work seems to be __. I think this photograph was taken in __. I feel it would be a __ place to live in because __.

New arrivals to a class

The greatest difficulty will be where there are recent arrivals in the class. In a secondary school the pupil may not be conceptually competent to deal with abstractions, but her language level may allow labelling of concrete objects and an understanding of simple requests. There may be opportunities to use bilingual education. If the pupil is literate in the home language it may be possible to use materials written in that language. In some cases materials in two languages may be available. There may also be bilingual people around who can translate for the student. These can include students in the class, teachers in the school, specially appointed bilingual teachers, parents, or community organization. If they cannot attend the lessons, they may be able to record something on a tape or in writing.

The availability of bilingual people and resources will vary considerably between the authorities, and the teacher may also feel unsure whether the concepts have been adequately explained in the other language. More significant for most teachers will be the learning of geography through English. What can the geography teacher do? First of all it is important to recognize that new arrivals to a country go through a long period of listening. The teacher may worry that nothing is being done in class. In fact the student is probably listening to conversations and learning the sounds of English. A range of early readers could be made available for general reading, chosen by the students and read at their own pace. At the same time some structured language work could be organized, using some of the easier methods already described. Ideally this should be co-ordinated with the language work done with English or ESL teachers.

LANGUAGE TO CHALLENGE RACISM

A critical look at language in textbooks and the media

The language of textbooks and resources can perpetuate racism through stereotyping and the use of problematic terminology, and through negative and excluding terms and phrases. Stereotyping and over-generalization ignore the rich complexity and diversity within cultural groups. Although they may result from a genuine desire to simplify knowledge to help children's understanding, they can become the basis of prejudice and discrimination. Care is also needed when using terms to describe or name different communities. The term 'race', for instance, is a problem. Although there are observable physical differences between people, most classifications of race are unsatisfactory because they attempt to impose discontinuous categories on a complex web of continuums, relegating many people to an unacceptable twilight zone between races. Names for groups can also cause difficulties. In multicultural education the names Native American, Inuit, and Traveller are used instead of Red Indian, Eskimo, and Gypsy. These changes are similar to the renaming of African countries after independence from colonial rule.

In Britain communities are often named according to their country of origin, although they are British. The term 'West Indian', for example, is an inaccurate description of black British Caribbean people, although some members of this community do not mind the term. Describing people as 'black' also leads to difficulties. It is sometimes restricted to those with 'Afro-Caribbean' connections, or it may be broadened to include the 'Asian' community, or other minorities such as the 'Chinese'. This broader use emphasizes 'black' as a political term, embracing groups with similar experiences of discrimination. Used in this way, the term 'black' becomes very controversial.

'Racism' is another difficult word. One meaning sees racism as personal prejudice and discrimination against black people. A second meaning emphasizes institutional racism. Streaming in school, for instance, may discriminate against black children, although the teachers involved are not prejudiced. A third meaning sees racism as a divisive social process, linked with colonialism and class divisions. Some members of the Irish community in Britain use the phrase 'anti-Irish racism', as found in anti-Irish jokes, which symbolize the colonial oppression of Ireland. With its strong negative connotations, variety of meanings, and disputed application, the term 'racism' can lead to heated debate.

Some descriptions are obviously negative and can be clearly labelled racist, e.g. describing groups as stupid, lazy, etc. Some words appear to be descriptive but have implicit values or emotional undertones. The word 'immigrant', for instance, is used inaccurately to describe black people and exclude them from being British. For this reason I have not used it in this chapter. Words like 'developed', 'civilized', and 'primitive' should not be used without discussion. 'Black' is more often used as a negative than a positive term, and teachers should become aware of the number of times they use it this way. This does not mean its negative use should be abolished. The aim is to achieve a better balance between positive and negative usage.

There are also subtler ways of using language to undermine ethnic minorities. One is constantly to present these groups as passive objects rather than as active subjects, through the syntax of the sentence. Another way is to use terms and phrases that deny their British identity, e.g. *'Their way of life contrasts with British culture.'* A good introduction to racism in children's books is to be found in Hicks (1981: 87-92).

Language values in education

One way of combating racism is to involve pupils in the process of values education. Pupils will need the opportunity to express opinions and explore controversial issues through reading, writing, and talking. Good advice on the use of discussion groups can be found in the *World Studies 8-13* Teacher's Handbook (Fisher and Hicks, 1985) and *People before places* (DEC, 1985), both useful books for geographers. Writing skills also have a significant role through the expression of attitudes, weighing of evidence, justification of choices, empathetic stories, etc. Also important is the audience for the writing. On the one hand it may be appropriate to allow confidential writing where pupils can express their feelings privately. On the other, it may be possible to broaden the audience outside the classroom, as in the production of a class newspaper. Using language this way is essential for challenging racism. Walford (1985) is a useful introduction to the main issues.

REFERENCES

Barnes, D. (1975) *From Communication to Curriculum*, Harmondsworth: Penguin.
Council on Interracial Books for Children Inc. (1977) *Unlearning 'Indian' Stereotypes*, New York: CIBC.
Crystal, D. (1984) *Who Cares about English Usage?* Harmondsworth: Penguin.
Dale, P.S. (1976) *Language Development*, London: Holt Rinehart & Winston.
Development Education Centre (1985) *People before Places?* Birmingham: DEC.
Dulay, H.C., et al. (1981) *Languages Two*, Oxford: Oxford University Press.
Dunsbee, T., and Ford, T. (1980) *Mark my Words*, London: Ward Lock.
Edwards, V. (1983) *Language in Multicultural Classrooms*, London: Batsford.
Fear, J., and Fear, S. (1985) *Apache Family*, London: A. & C. Black.
Fisher, S., and Hicks, D. (1985) *World Studies, 8-13*, Edinburgh: Oliver & Boyd.
Hicks, D. (1981) *Minorities*, London: Heinemann.
Houlton, D. (1985) *All our Languages*, London: Arnold.
Inner London Education Authority (1981) *The World in the City*, London: ILEA.
Jones, A., and George, N. (1981) 'The subject teacher in a multicultural school', in C. Sutton (ed.) *Communicating in the Classroom*, London: Hodder & Stoughton.
Kress. G. (1982) *Learning to Write*, London: Routledge & Kegan Paul.
Levine, J. (1982) 'Developing pedagogies for multilingual classes', *English in Education* 15, 3: 25-33.
Lewis, D. (1987) 'What do your pupils write?', *Teaching Geography* 12, 2: 60-3.
Lunzer, E., and Gardner, K. (1984) *Learning from the Written Word*, Edinburgh: Oliver & Boyd.
Naidoo, B. (1985) *Journey to Jo'burg*, London: Longman.
Preusch, D., et al. (1981) *Red Ribbons for Emma*, Stanford, Cal.: New Seed Press.
Raleigh, M. (1981) *The Languages Book*, London: ILEA.
Richmond, J. (1979) 'Dialect features in mainstream school writing', *New Approaches to Multiracial Education* 8, 1: 9-15.
 (1986) 'What we need when we write', *Newsletter* No. 2 of the National Writing Project, London: SCDC.

Riley, S. (1986) 'The Second Language Learners in the Mainstream Classroom Project', *Multi-ethnic Education Review*, 5, 1:21-2.

Scott, S. (1985) 'Collaborative learning', Second Language Learners in the Mainstream Classroom Project, Paper No. 3.

Searle, C. (1983), 'A common language', *Race and Class* 25: 65-74.

Slater, F. (1982) *Learning through Geography*, London: Heinemann.

Sutton, S. (1981) *Communicating in the Classroom*, London: Hodder & Stoughton.

Trudgill, P. (1975) *Accent, Dialect and the School*, London: Arnold.

Walford, R. (ed.) (1985) *Geographical Education for a Multicultural Society*, Sheffield: Geographical Association.

Part II

LANGUAGE AND CONCEPT DEVELOPMENT

Chapter Six

CONCEPT MAPS AND CHILDREN'S THINKING: A CONSTRUCTIVIST APPROACH

Anthony L. Ghaye and Elizabeth G. Robinson

INTRODUCTION

This chapter argues that there are sound educational reasons why a central concern for teachers might usefully be the development of methods of enquiry that enable them to discover and understand their children's 'structures of thought'. In support of this argument it presents some of the more significant insights of two teachers engaged in such work.

The evidence for this chapter has been gathered periodically across six years during our work with practitioners and children between the ages of 5 and 16 years. Sometimes we have been in the role of class teacher engaging in small-scale action research enterprises, while at others as interested outsiders facilitating classroom-based or whole-school enquiries.

The notion of pupil structures of thought in the context of the natural classroom setting begs a number of conceptual and procedural questions. For example, we need to establish some shared understanding about the nature of knowledge structures in terms of their different qualities. We need to know something about the cognitive processes associated with the act of building or constructing meanings. We also need to know how the issues of the teaching and learning of knowledge structures can be usefully 'framed' so that busy teachers may explore them in an effective manner. A critical part of the generation of insights and understandings with regard to the issues of the building and reconstruction of meaning is a recognition by teachers of the value of the deconstruction of the familiar and the importance of thoughtful, reflective collegial reconstruction. It is this process that has helped all those involved in our work to achieve new levels of awareness, competence, and confidence. One of the central purposes of this chapter is to describe some of the important aspects of this deconstruction-reconstruction process and to present some of the understandings and sensitivities that we acquired through a committed and sustained engagement in this process.

CLARIFYING THE CONTEXT

To properly contextualize what is to follow and therefore to get some purchase upon the main concern of this chapter, we need to explore, if only briefly, the dimensionality of the term 'structure'. Structure in the way it is used in this chapter is a vast *terra incognita* for many practising teachers. There are a number of reasons for this. First, the sub-concepts of knowledge and cognitive structures are rarely addressed in any sustained and systematic manner in pre-service or in-service teacher education courses in the United Kingdom. It should come as no surprise therefore that there is a basic problem for teachers of knowing how to talk about these concepts, how to use them to identify and to organize curriculum content and inform their teaching. Some teachers, for very plausible reasons, are wedded to a relatively narrow and limiting conception of structure that embraces only the notions of time and order. In addition there are methodological problems with regard to the way teachers might elicit, and children represent, in a meaningful way, their acquired and developing cognitive structures. Both teachers and children then need to be able to recognize the qualities inherent in any legitimate representation of part of their cognitive structure and to be able to discuss these qualities. There is also some uncertainty among teachers associated with the educative value of a 'constructivist approach' to teaching and learning, one which emphasizes the idea that the learner has to construct and reconstruct meaning in the light of past experiences, current preoccupations, and their future intentions. In part this uncertainty arises from a lack of evidence, pointing to the value of such an approach, that can easily be accessed and understood by teachers. Without evidence of this kind it is hard to see how teachers can make any kind of valid judgement about how understanding pupil structures of thought might help them teach more effectively and their children learn in a more efficient and meaningful way.

Schwab (1962) and Bruner (1960) have both provided clear and persuasive arguments as to why practitioners should be aware of and interested in structure. Schwab's position is centred upon the idea that any discourse about structure needs to point up its conceptual and syntactical components, and he goes on to argue that considerations and analyses of this kind help us to ask more fundamental questions about how far and in what ways curriculum content is warranted and meaningful. Bruner's discussion of structure also serves this chapter well, particularly his point that 'Grasping the structure of a subject is understanding it in a way that permits many things to be related to it meaningfully. To learn structure in short is to learn how things are related' (1960:7). This emphasis upon the interrelationships between ideas and procedures and the need to make them explicit and comprehensible to pupils is regarded by us as critical and central to the process of teaching and learning.

In this chapter we draw a distinction between knowledge structures and cognitive structures. The former are taken to mean something associated with the tasks of general curriculum or specific lesson planning and instruction. An example is shown in figure 6.1, which is a simple structure for a series of lessons on the topic of water, taught by one of us to a class of thirty 11-12 year olds in a comprehensive school. It shows the most important ideas and relationships that Tony wanted to teach and that he hoped the children would grasp. It is a basic framework of meanings which could be changed, extended, and filled in by Tony and the children. Cognitive structure on the other hand refers to the way children, under certain situational and task-related influences, organize and reconstruct these knowledge structures. Although there is a lack of definitional consensus about cognitive structure, we felt that this definition was both adequate and appropriate for the purpose of our work in school. A person's cognitive structure, even for a relatively small content domain, can be extremely complex. Cognitive structures are also unique (Van Den Daele, 1969) and differ from each other with regard to particular organizational characteristics. We attempt, latterly, in this chapter, to describe some of those characteristics made manifest through the medium of 'concept maps' and offer some suggestions as to why children, with certain kinds of ability and intention, organize their expanding knowledge base in the way they do.

We believe that an important part of the teacher's role is to try to understand how each child's mind works, what makes it 'tick', and what changes when learning occurs. Cognitive structures change, but we wanted to know in what general kinds of way. We also wanted to address more explicitly and squarely some of our assumptions with regard to the links between the to-be-learned subject matter (the content of each

Figure 6.1 A simple knowledge structure diagram for a series of lessons, to a class of 11-12 year-old children, on the topic of water

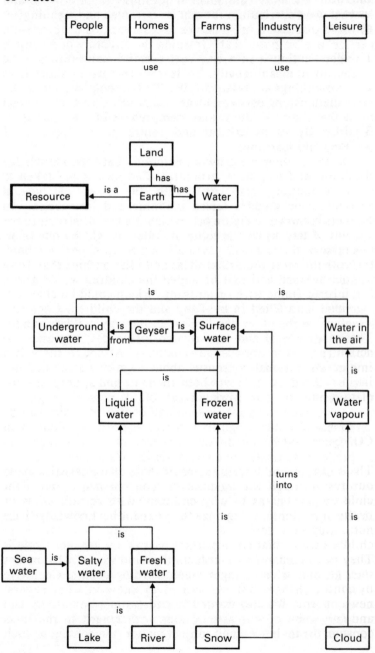

Concept maps and children's thinking

geography lesson) and the nature, pace, and direction of changes in cognitive structure complexity. The evidence, in the form of children's concept maps, that emerged encouraged us to think more deeply about questions and assumptions such as these:

1 When and how far should there be a match between what is taught and learned?
2 How far we assume that orderliness and rationality are salient characteristics of a child's developing conception of a body of geographical knowledge.
3 How far we proceed on the basis that the variety of tasks the children engage in have the same or differential effects on their cognitive structure.
4 In what ways a child's motivation, general attitude to school, approach to learning and self-concept, affects the child's meaning-making process.
5 Do changes in a child's cognitive structure proceed in some uniform manner? Which parts of the content domain-related structure extend, grow, decay, are held in abeyance and drawn upon from time to time, or become quietly obsolete as learning occurs? Do some kinds of link between informational items establish themselves before others? Is there a sequence, and so on?

These are challenging questions that call teachers' theories-of-action (Argyris and Schon, 1974) into question, call certainty into question, challenge conventional practitioner wisdom, and invite teachers to deconstruct some habitual and routine ways of thinking about and teaching subject matter.

CONCEPT MAPS

The children's labelled-line concept maps that were used in our research are a kind of construction technique, in that children are invited to establish and build relationships between concepts in ways that make sense to them. They were not used just as a way of discovering what or how much the children were able to recall about a lesson or series of lessons. They were used to find out how the children were structuring their geographical experiences, how the structuring was biased by context, how ideas and relations got deleted or altered, how new concepts and links between them were added to fill gaps and forge new relations, how they represented the children's constant struggle to make sense of what was being taught so

Figure 6.2 Russell's labelled-line concept map

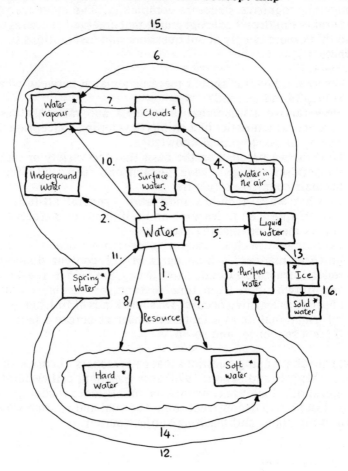

1. is a common
2. Sometimes is
3. is sometimes
4. is really called clouds
5. that we use is
6. is also called
7. is called
8. is sometimes
9. is sometimes
10. sometimes turns to
11. is very clear purified
12. is
13. is not
14. is very
15. is
16. is

Figure 6.3 Christopher's labelled-line concept map

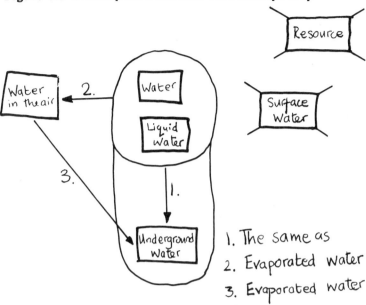

as to develop for themselves a coherent body of knowledge. Concept maps are therefore related to a network model of semantic memory which consists of nodes (concepts, ideas) and links (labelled relations between nodes). Our work also reflects the Martonian view (Marton et al., 1984) that:

> The learners construction of meaning (or development of a conception) of content is the very heart of the learning experience. We consider the finding and describing of conceptions (meanings) of fundamental aspects of various learning materials to be one of the main tasks of research into student learning. (Marton and Saljo, 1976:5)

Figures 6.2-3 provide two illustrations of what labelled-line concept maps can look like, and should be seen in relation to figure 6.1. Both Christopher and Russell had attended all the lessons on the topic of water. Both boys had been given the same invitation to build a concept map around the six concept cues of 'water', 'liquid water', 'underground water', 'water in the air', 'surface water', and 'resource'. Both were given up to forty minutes' on-task time. This was not the first occasion that the boys had been asked to undertake this type of activity. Evidence of this kind intrigued us and underlined for us the importance of trying to understand the interactions

Figure 6.4 Organizational framework for each lesson

Lesson No. 25

                Organizing concept   **Spatial interaction**

Sequence of key questions

1. What is the Country Code?

2. Why are people encouraged to follow the Country Code?

3. What is the role of the Countryside Commission?

4. In what ways is the countryside a resource?

Main resources

1. A route through the countryside in the form of a series of numbered squares, with ten cards each containing one rule from the Country Code.

Pupil task(s)

1. Pupils move along the route, using a table of random numbers, and follow the instructions on each square upon which they land.

2. Pupils design an eye-catching poster for the Countryside Commission.

3. Pupils judge the merits of each other's poster.

Knowledge structure diagram

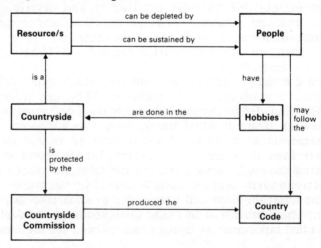

between prior knowledge and novel information, between a child's current frame of mind and future intention, between semantic and episodic knowledge.

PROCEDURES AND THEIR IMPLICATIONS

The teacher

A number of techniques for representing the structures in knowledge together with a justification of their use are available (Lewis and Woolfenden, 1969; Buzan, 1973; Novak, 1977; Weil and Joyce, 1978; Hawkridge, 1981). Reassuringly Rowntree suggests that:

> many people setting out to draw a concept map get obsessed with uncovering the 'true underlying structure' of the subject. Even if such a structure exists, there is no need for that degree of dedication. Drawing such diagrams is a personal and subjective activity whose purpose is simply to help you externalise your own understanding about how the concepts within your field connect up. So long as you find it helpful in your own course planning, there is no need to worry that others might have mapped the territory differently. (1982:105)

Thus teachers have to think about their lessons in terms of key ideas or concepts, which are the fundamental agents of intellectual work, and the most important, valid, relevant, or most appropriate relationships that exist between them, given, for example, the nature of the content domain, their aims and objectives, and the children's abilities, predispositions, aptitudes, needs, and wants. At the preactive stage, then, the teacher needs to construct a knowledge structure diagram which acts as the core of their lesson plan. Figure 6.4 shows the kind of plan we used in much of our work.

The knowledge structure is a spatially organized visual framework. The key ideas are shown in boxes together with the semantic or meaning relations between them that the teacher intends to explore and discuss with the children. Building knowledge structures of this kind is a thoughtful and reflective activity. For example, teachers have to erect criteria for the selection of concepts and the important links between them. They have to focus on what kinds of procedures, individual tasks, and group activities will be necessary for the teaching and comprehension of a particular structure. They have to decide how they will draw upon and relate to the knowledge structure during the course of the lesson. Teachers have to think about the things that give the structure coherence, how it is related to previous lesson plans with other knowledge structures, what potential the knowledge structure has for extension and elaboration, and its potency for enhancing more meaningful pupil learning.

Teachers should not feel that the kind of constructivist approach described here means that they have to adopt a transmission style of teaching and that the children have to regurgitate the teacher's knowledge structure. Children should not feel that knowing is equated with remembering the structure that the teacher has already organized for them and that rewards and confidence depend upon how far their restatement of the corpus, their labelled-line concept map, corresponds to some authority figure's statement, namely the teacher's knowledge structure. We believe that an important attribute of learning is the building and rebuilding of structures (Renner and Lawson, 1973). It is incumbent upon practitioners therefore to devise structures that do not restrict, distort, and confine learning but enable children to go further more easily. Conveying a 'sense of problem' (Popper, 1972), which is such a crucial ingredient in teaching, is lost when children slavishly follow rigid structures. Good teachers appreciate this. They also appreciate the need to help children to structure, enrich, and reconstruct their own experiences. Only then are children really ready to study, critically, the structures of others.

The child

The instructions to the child must be simple and straightforward. With older ones we found ourselves using the technical term 'concept maps'. With the younger children we called it a sticky dot exercise.

A number of things have to be considered before children are asked to construct a concept map. First the teacher has to think about the purpose of the task and then create an appropriate context for children to build their maps. Without doubt, evidence in the form of concept maps provides some very useful diagnostic data for the teacher. But concept maps need not be restricted to this role. The mapping technique can play an active part in the learning process itself. Constructing concept maps is a worthwhile pupil activity in its own right.

It is a kind of problem-solving task. It invites the children to learn the skills of how to handle, organize, and represent their expanding knowledge about something. It gives children the freedom and responsibility to learn what to put into the map, to select the most important or appropriate ideas. They have to learn where to put them and what can justifiably be linked with what. The maps make what the child has learned more visible. Therefore children should be encouraged to see the task as one where they can inspect and reflect upon their existing frame of mind more easily.

> before starting a network, one often has the feeling of possessing a large, rich and sensitive, albeit not quite clearly formulated set of thoughts ... but when one does come to write it out as a network, what strikes one most at first is its paucity, naivety and inner contradictions. At the least this can act as a strong stimulus to do better ... (Bliss et al., 1983)

The teacher also has to think about when to ask the children to undertake the task. On the basis of this decision the teacher should then analyse the knowledge structure diagram(s) that formed the basis of the lesson or scheme of work. A limited number of concepts then need to be chosen to act in a catalytic role. The selection of these concepts is problematic, and the teacher needs to erect a robust set of criteria to guide the choice. For children older than nine years, we found that six concepts were the most appropriate number to present. Normally these would be written on sticky labels, and each child would then receive a blank sheet of paper, a set of six concepts, and a list of instructions such as the ones set out below:

1. Look at the sheet of sticky labels. Read each word on each label.
2. Choose which label you want to peel off first. Peel it off and stick it down.
3. Choose a second label. Peel it off and stick it down. Try to link the two words together. Draw a line to do this. It does not have to be straight.
4. Draw as many lines as you like. Write on each line to show how you are linking your sticky labels together. Another way to do this is to put a number on each line and on the back of your paper write the words to tell how you are linking labels together.
5. Now do this again using all the other labels. Stick them

down and link them up in a way that makes most sense to you. The idea is to build up a picture of labels and lines.
6 If you think some of the labels mean the same thing, then stick them down and draw a line around them.
7 If you do not know what a label means, stick it down and cross it out.
8 You can use as many labels as you like. These six are just to start you off. You can draw as many labelled lines as you like.
9 Don't rush. Think.

The concept map task was done individually but there is plenty of scope for the teacher to set up a context and write some instructions that turn it into a collaborative group exercise. In addition the teacher does not have to provide six concept cues in the form of words on sticky paper. Depending upon the class, a teacher may wish to begin with pictures on cards rather than words on paper. We have started this way with 5-year-old children when exploring their perceptions of the links between 'home' and 'school'. More pictures were then added. We used ones to represent 'teacher' and 'parent', for example. A mix of words with pictures can then be used until the six-concept cue situation is reached. This kind of progressive development of stimuli that help children to build meanings may span a number of terms or years. Children who have some difficulty in expressing their ideas in writing should be encouraged to talk about the links on their maps with the teacher. At times we found this rather time-consuming. As an alternative strategy we provided groups of children with a tape recorder and an invitation to talk about their constructions through that medium.

SO WHAT DID WE LEARN?

Much of the evidence for this chapter has been gathered over six years and so the insights and understandings that have emerged are substantial and varied. Space only permits us to focus upon some of those things that might have the most immediate relevance for classroom practice. We hope that what we present is intelligible, plausible, and practical. Table 6.1 serves to bring together and convey something of the flavour of what we have learned, portrays some of the main characteristics of the child's cognitive structure as represented by the concept map, and suggests some of the reasons why children display and arrange their knowledge in the way they do. It is offered as a framework to guide the teacher's thinking about a constructivist approach in relation to certain categories of learner. The main task of the teacher is to enable the child to move from right to left on the diagram. The pedagogic implications need to be thought through carefully. In addition to being a heuristic device we also hope that table 6.1 serves an emancipatory function in drawing the teacher's and child's attention to what can be changed, what does change, and why.

Complexity of concept maps and knowledge structure diagrams

This is the easiest characteristic for teachers to appraise. Complexity refers to the amount of structure shown on a map. Simply put, it is defined as the relative number of appropriate links to concepts. Robinson (1986) developed a useful way of representing complexity. It is shown in figure 6.5. She took the number of appropriate links and concepts as co-ordinates and plotted the position of knowledge structures and concept maps on a graph.

Graphs such as this could be kept for each child over a period of time. A profile of changes in concept map complexity could then be seen quickly and easily. Individual graphs would need to be properly contextualized by seeing them in relation to the class portrait and the complexity of the teacher's knowledge structures.

Concept map complexity represents the interplay between the connectedness of the teacher's knowledge structure and the way in which the child has stored, organized, and then been able to recall and reconstruct meaning. We found that the intervening time between the lesson(s) and the concept map had some effect on complexity. It reduced with time. However, Ghaye (1984) found that in time complexity did reduce, and did so to a point where, on average, a child's concept map was half as complex as the teacher's knowledge structure. Quality did not decay and usually increased. Variety increased also.

With help from the teacher, and with experience that comes from practice, the complexity of a child's concept map can be increased. Most children understand more than is shown on their map. Some argue that the less able, younger, less conversant learners (Paris and Upton, 1976) produce skeletal representations because they are unable to remember enough information to perform integrative and inferential

Table 6.1 Concept maps and children's thinking: towards a typology of reconstructions

Demonstrated by the more able child, more experienced learner, or those more conversant with the content domain	Characteristics	*Represented most often by the less able child, less experienced learner, or those less conversant with the content domain*
Complete, cohesive, complex representations, multi-structural, interconnected, ability to make inferences generates an elaborate map	COMPLEXITY	Disjointed, simple, skeletal representations, high degree of information is redundant, linear patterns, unstructural
High proportion of links between concepts are appropriate, valid, acceptable. Ability to use 'high order' links (e.g. logical)	QUALITY	High proportion of links between concepts are inappropriate, confusing, erroneous. Predisposition to a sustained use of 'lower order' links (e.g. structural)
Maps show a variety of kinds of link, imaginative use of new links	VARIETY	Similarity, limited number of different kinds of link used.
Mismatch with the to-be-learned content, imaginative, eclectic, divergence, building up of a personalized semantic structure	MATCH	A close match with the to-be-learned content, replication of what was taught, convergence
Changes over time, reflects interaction with new knowledge structures, dynamic, reorganization	DYNAMISM	Relatively stable over time, sameness, static, reproduction
Deep approach, getting the gist, personalizing the teacher's knowledge structure, reconstruction rather than reproduction	APPROACH	Surface approach, learning by rote, trying to memorize the teacher's knowledge structure, preoccupation with reproduction

Figure 6.5 Plotting concept-map complexity on a graph. x position of concept map on graph. Area A: more complex concept maps. Area B: less complex concept maps

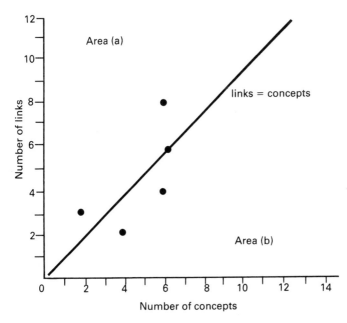

operations. Our own feeling is that these children may well be able to integrate and infer and thus produce quite complex, interconnected maps. A major problem is that they are often unaware that they need to do so in order to increase their ability to remember and understand. In classrooms where a constructivist approach is adopted, teachers are alive to those comprehension-monitoring skills, encoding and processing strategies, and retrieval mechanisms that help to increase the complexity of the child's cognitive structure and their ability to represent this complexity meaningfully through the medium of the concept map.

> Understanding depends upon being able to develop a web of interconnections which relate previous knowledge and experience to the new ideas being represented. All too often it seems that knowledge in school is presented in ways which make it difficult for pupils to make necessary connections particularly with their own experiences in everyday life. (Entwistle, 1987:37)

Figure 6.6 Classification of links between concepts

Types of relationship	Some of the more frequently occurring expressions of the relationship	Some exemplars of concepts and relationships from pupil concept maps	
1 DESCRIPTIVE-TYPE RELATIONSHIPS *Structural kind* expresses a taxonomic or hierarchical relationship between concepts such as sub-set, inclusion, co-ordinate, 'kind' or 'parts' relationship	is a (kind of) has/have means is a sort of concerns includes	**Liquid water** **Dredging** **Area for recreation** **Coal**	is a is a kind of is made up of is known as	Resource Mining Places for recreation Fuel
Functional kind relationships expressing a function, purpose or use	needed for used by/for for are useful to gives uses	**People** **Water** **Maps** **Gas**	use is needed for are useful for finding is used for	Resources Living Locations Heat
Locational kind are relationships that locate concepts in space	is from is below is found in on under north/south/east/west of	**Coal** **Oil** **London** **Tin**	lies is below is south-east of is found in	Underground Earth's crust Worcester Malaysia

2 DEPENDENT-TYPE RELATIONSHIPS

Procedural kind
express an order, sequence of steps, progression, precondition, process or prerequisite relationship between concepts

	turns into	Surface water	evaporates into	Water in the air
	is got by	Iron ore	is extracted by	Open-cast mining
	first, then, makes			
	extracts			
	done	Living things	eat	Dead things

Logical kind
expresses a logical or conditional relationship between concepts

	is sometimes	Resources	can be	Depleted
	can be	Water	is sometimes	Hard water
	may follow	People	may follow	Country code
	might disturb			
	could be			
	is mostly	Living things	are the opposite to	Dead things

3 COMPOSITE-TYPE RELATIONSHIPS

Drilling	is done on	Sea	and around the	British coast	and on Dry land
we use a	Scale	to find the	Distance
Man	can	Extract	Resources

4 ERRONEOUS-TYPE RELATIONSHIPS

Renewable resource	you can't get rid of	Gold		
Site for recreation	is a certain	Distance		

The quality and variety of links between concepts

A number of studies have tested the assertion that representing aspects of the learner's cognitive structure is feasible in the natural setting (Shavelson, 1971; Fenker, 1975; Rudnitsky, 1976; Boekaerts, 1979; Champagne *et al.*, 1981; Stuart, 1983). There has also been some interesting work based upon the assumption that the meaning of each concept is a function of the set of meaning relations which hold between the concept and other concepts in the same 'domain', thus pointing up the need to label the links between the words on the sticky labels (Frijda, 1972; Stewart, 1979, 1980). However, there have been only a relatively small number of attempts to set out a procedure and to classify the kinds of relationship evidenced on children's concept maps (Dyer, 1969; Bliss *et al.*, 1983; Ghaye, 1984; Robinson, 1986). For teachers who plan to devise a classification we suggest the following two-step procedure:

1. Take each knowledge structure diagram. Analyse the nature of the links between the concepts. Record them as being of certain kinds, e.g. functional, locational, casual, temporal, logical.
2. Take each concept map and test the comprehensiveness of the classification and definition of each kind of link that has emerged from 1 above.

The classification one of us developed (Ghaye, 1984), related to one year's work based on the theme of 'resources' and organized by the concepts of 'scale', 'location', 'spatial interaction', and 'spatial change through time', for a class of 11-12 year olds, is shown in figure 6.6.

To a certain extent the kinds of links identified will vary between content domains. But trying to classify the links in some kind of way that makes sense to all the participants is the important thing. It is a bonus if a teacher's classification is sufficiently elaborate to be transcontextual and transdisciplinary. The outcome of this procedure is that it enables teacher and child to make meaningful comparisons between what was taught and the sense the child has made of it. Figure 6.7 shows how the classification shown in figure 6.6 has been used to analyse one child's concept map.

Concept maps and children's thinking

Figure 6.7 Maria's labelled-line concept map (redrawn and reduced)

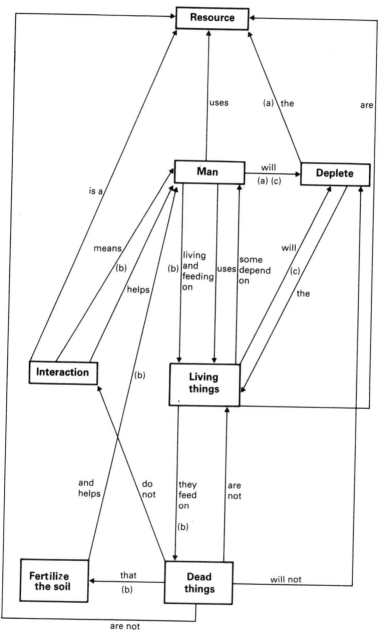

Table 6.2 An analysis of Maria's concept map, figure 6.7

	Descriptive				Dependent			Comp. (%)	Err. (%)	Complexity
	St.	Func.	Loc.	Total (%)	Pro.	Log.	Total (%)			
Maria's map	10	10		20		15	15	45	20	20
Class portrait	18	15	1	34	7	8	15	37	13	12
Teacher's knowledge structure	33	5		38	48	14	62	–	–	21

Matching what is taught with what is reconstructed

Devising classifications of this kind makes substantial demands on a teacher's time and powers of analysis. However, once it has been done, making modifications is easy. Once it is done, teachers are in a position to reap rich rewards for their efforts. For clarity we have redrawn Maria's concept map (see figure 6.7) for the reader. Maria was one of the most able and articulate 12 year olds in the class. The analysis shown in table 6.2 says something about the general quality of her map and the repertoire or variety of links Maria has drawn upon to structure her thoughts. But when the individual case is put against a class portrait and, further, set against a similar analysis of the teacher's knowledge structure diagrams that comprised the series of lessons, a rich picture emerges of the match or ˙ congruence between what was taught and reconstructed.

The static-dynamic dimension

Although we have discussed in some detail the benefits of such comparisons elsewhere (Ghaye, 1986; Ghaye and Robinson, 1987) it is worth stressing the following at this point. We found that when children made inferences, engaged in 'slot-filling' or 'text-connecting' strategies (Warren *et al.*, 1979), if they personalized their map, the match or correspondence with the teacher's knowledge structure was low. We became attracted to the idea that match-mismatch was an indicator of change, something to signal whether the processes of construction and reconstruction were being activated by the learner. A close match with the teacher's knowledge structure may often be indicative of a more passive and reproductive learning posture rather than a meaning orientation. There is some interesting research that supports this suggestion (Rudnitsky, 1976; Barrett, 1985).

Approach to the task

Learning is a complex business. An important aspect of this enterprise is the learner's intention. What they intend to get out of the concept mapping determines, to a great extent, their approach to it, their manner of proceeding. This in turn determines the quality of the observed learning outcomes. We have deliberately left a discussion of this aspect of table 6.1

until now because we wanted to conclude the chapter by underlining the fact that our intention of 'concept maps and children's thinking' is essentially for it to serve a heuristic and emancipatory function. This chapter merely represents the finding of a key to unlock a door to a vast and intriguing domain.

In general we believe that behaviour is context-dependent. Therefore it is plausible and permissible to suggest that a child's approach to the concept-mapping task is situation-determined and task-specific. However, our evidence points to the fact that the more able, more experienced, more conversant learner was able to choose between a deep or surface approach to the task, whereas the other categories of learner showed a more sustained predisposition towards a surface approach only. However, we do not want to give the impression that a child's approach, which plays such an important part in the quality of the concept map, is determined only by cognitive factors. The organizing effect of the child's prior knowledge, current purposes, and future intentions is also influential.

Much more needs to be said about the links between the characteristics shown in table 6.1 and the child's perception of the task. A possible agenda would include a discussion of the importance children accord the task and its context, their feelings about how well they can cope with it, what they can get out of it - all of which are closely linked with their interest in it.

SOME FINAL THOUGHTS

This chapter covers fascinating territory. It has endeavoured to present some insights and understandings that have accumulated over six years as our interest has grown in the issue of building meanings and in the deconstruction-reconstruction process that plays such a crucial part in the relationship between what is taught and learned. Implicit in our discussion is the fundamental question of the relationship between evidence and theory. How do you see this? By using concept maps and devising the procedures outlined in this chapter we feel that we have moved some way towards establishing a tentative 'typology of reconstructions'. Even so, it also feels as if everything still needs exploring and discussing. Much more work clearly needs to be done. Your reactions to this chapter will influence what we do next.

This procedure helps us to look beyond a child's

knowledge of a topic or theme. Rather than saying that a child knows about the topic of water, or the theme of resources, we are moving towards a position where we can say that they understand classifications or that they can organize their knowledge in a logical manner, that they can think in casual or functional terms, and so on.

Nisbet and Shucksmith (1986) suggest that teachers should make every effort to help the child develop a range of metacognitive skills. They refer to this as the underdeveloped seventh sense among many people generally (Nisbet and Shucksmith 1984). Metacognition is the ability to reflect on how one learns and ultimately perhaps the ability to respond to the demands of the task-in-context more appropriately and richly. The procedures of analysis and comparison presented here help to develop these metacognitive skills in the learner. For example, if the teacher feels that the majority of relationships between the ideas she is teaching are 'structural' in kind she could help pupils to learn by asking them to look for similarities or categories in the to-be-learned content. On the other hand if children knew that a teacher was structuring a lesson using 'procedural' links, they might come to appreciate that their learning could be more efficient and effective if they drew upon the kind of thinking that consisted of steps or contingencies. The approach outlined in this chapter not only challenges teachers to come to know how the child understands and makes sense of the lesson content. It also challenges practitioners to reflect upon their own teaching and understanding of the subject matter.

REFERENCES

Argyris, C., and Schon, D. (1974) *Theory in Practice*, San Francisco: Jossey-Bass.
Barrett, G. (1985) 'Structure of knowledge and the learner: an examination of cognitive learning skills, learning situations and knowledge', *Cambridge Journal of Education* 15, 2:73-80.
Bliss, J., Monk, M., and Ogborn, J. (1983) *Qualitative Data Analysis for Educational Research*, Beckenham: Croom Helm.
Boekaerts, M. (1979) 'Can the teacher unravel parts of a pupil's cognitive structure?' *Journal of Experimental Education* 48, 1:4-17.
Bruner, J.S. (1960) *The Process of Education*, Cambridge, Mass: Harvard University Press.

Buzan, T. (1973) *Use your Head*, London: BBC Publications.

Champagne, A., Klopfer, D., Desena, A., and Squires, D. (1981) 'Structural representation of students' knowledge before and after science instruction', *Journal of Research in Science Teaching*, 18, 2: 97-111.

Dyer, J. (1969) 'The teaching and assessment of cognitive structure through the diagrammatic representation of structures of knowledge', unpublished Ph.D. thesis, Michigan State University.

Entwistle, N. (1987) *Understanding Classroom Learning*, London: Hodder & Stoughton.

Fenker, R. (1975) 'The organisation of conceptual materials: a methodology for measuring ideal and actual cognitive structures', *Instructional Science* 4: 33-57.

Frijda, N. (1972) 'Simulations of human long-term memory', *Psychological Bulletin* 77: 1-31.

Ghaye, A. (1984) 'Discovering geographical mindscapes: a participant observation study of children in the middle years of schooling', unpublished Ph.D. thesis, University of London, Institute of Education.

Ghaye, A. (1986) 'Discovering classroom underlife', *C.A.R.N. Bulletin* 7: 232-41.

Ghaye, A., and Robinson, E. (1987) 'Discovering through discourse: topic teaching and learning outcomes', *C.A.R.N. Bulletin* 8.

Hawkridge, D. (1981) 'The telesis of educational technology', *British Journal of Educational Technology* 12, 1:4-18.

Lewis, B., and Woolfenden, P. (1969) *Algorithms and Logical Trees: a Self-instructional Course*, Cambridge: Algorithm Press.

Marton, F. (forthcoming) 'Phenomenography: a research approach to investigating different understandings of reality', in Fetterman, D. (ed.), *A Shift in Allegiance: the Use of Qualitative Data and its Relevance for Policy*.

Marton, F., and Saljo, R. (1976) 'On qualitative differences in learning', *British Journal of Educational Psychology*, 46: 4-11.

Marton, F., Hounsell, D., and Entwistle, N. (1984) *The Experience of Learning*, Edinburgh: Scottish Academic Press.

Nisbet, J., and Shucksmith, J. (1984) 'The seventh sense', *Scottish Educational Review* 16: 75-87.

(1986) *Learning Strategies*, London: Routledge & Kegan Paul.

Novak, J. (1977) *A Theory of Education*, Ithaca, N.Y.: Cornell University Press.

Paris, S., and Upton, L. (1976) 'Children's memory for inferential relations in prose', *Child Development* 47: 600-68.
Popper, K. (1972) *Objective Knowledge*, Oxford: Oxford University Press.
Renner, J., and Lawson, A. (1973) 'Piagetian theory and instruction in physics', *Physics Teacher* 11: 165-9.
Robinson, E. (1986) 'The teaching and learning of geographical structures: a small-scale investigation', unpublished M.A. thesis, University of London, Institute of Education.
Rowntree, D. (ed.) (1982) *Educational Technology in Curriculum Development*, New York: Harper & Row.
Rudnitsky, A. (1976) 'Content structure, cognitive structure and their relationship: a methodological investigation', unpublished Ph.D. thesis, Cornell University.
Schwab, J. (1962) 'The concept of the structure of a discipline', *Educational Record* 43: 197-205.
Shavelson, R. (1971) 'Some aspects of the relationship between content structure and cognitive structure in physics instruction', unpublished Ph.D. thesis, Stanford University.
Stewart, J. (1979) 'Content and cognitive structure: critique of assessment and representation techniques used by science education researchers', *Science Education* 63, 3:395-405.
Stewart, J. (1980) 'Techniques for assessing and representing information in cognitive structure', *Science Education* 64, 2: 223-35.
Stuart, H. (1983) 'Studies in cognitive structure: exploring secondary school pupils' understanding of genetics', unpublished Ph.D. thesis, Centre for Science and Mathematics Education, Chelsea College, University of London.
Van Den Daele, L. (1969) 'Qualitative models in developmental analysis', *Developmental Psychology* 1: 303-10.
Warren, W., Nicholas, D., and Trabasso, T. (1979) 'Event chains and inferences in understanding narratives', in R. Freedle (ed.) (1979) *New Directions in Discourse Processing*, Norwood, N.J.: Ablex.
Weil, M., and Joyce, B. (1978) *Information Processing Models of Teaching*, Englewood Cliffs, N.J.: Prentice-Hall.

ACKNOWLEDGEMENTS

Our special thanks go to the children and staff of the Eastbank, Marlpool, and Cox Green Schools for their helpful comments and support over the years with our work in this field.

Chapter Seven

THE FEELING TONE OF CHILDHOOD: CHILDREN WRITING ABOUT THEIR ENVIRONMENT

Ann Jones

INTRODUCTION

This chapter presents a collection of children's writings about place, in order to draw attention to the emphases, understandings, and feelings which the writing reveals about their responses to their environment. As well as noting the substance of their writing, the chapter includes descriptions of everyday encounters with ordinary places, and, as well as documenting the humdrum, their writing reveals the intensity and significance of certain environmental experiences. It is interesting to register the fluency and feeling tone of their writing of their experiences.

The essays were written by primary-school children as part of a study examining childhood responses to the landscape. The selection here includes examples from boys and girls in all age groups between seven and eleven years. I have corrected the spelling but kept the sentence structure intact to retain the vitality with which the pupils expressed their feelings.

Broadoak School is in Swinton, a suburb of Manchester. Most of the children lived in small estates of semi-detached houses and had access to a back garden. They also played in the open spaces and woods which have evaded urban development. A stream, threading through the clay hills, has cut a series of steep-sided hollows called 'dingles'. Unsuitable for housing, they remain, thickly wooded, as pockets of wilderness.

THE STRUCTURE OF EVERYDAY LIFE

The children's writing declares the gentle tedium of everyday life.

> *David*. I like playing on my road. I think it is very boring but it is good for playing things. I like it only a bit.

Their lives are constrained by school and the general demands of home. 'A lot of outdoor life for children adds up from bits.' According to Jane Jacobs (1961:96), snatched episodes of play form punctuation between schooltime and bedtime. Their outdoor lives are interrupted by piano lessons, gymnastic classes, and rain. This patchwork quality of everyday life is clearly displayed in John's writing:

> *John*. I play in many places after school. I often go to Max's house and there we go ghost hunting in his loft. If I feel like it, I go to my Nana's on my bike. I also play on the golf links with my next-door neighbour. At weekends I sometimes stay at my Nan's. Otherwise I stay at home, watch TV, or ride my bike. At weekends I also do gardening or help out my Mum. The activities I do range from football to biology, otherwise I do not do much. When it is raining I go into my bedroom and do biology, because I would like to be a biologist when I grow up.

Much of John's life seems haphazard. He drifts off to the golf course or goes to find a friend. Like most children in the study, he has obligations in the home, although he does not have to spend his time baby-sitting or taking responsibility for younger ones, as many others did. The child belongs to a family, and the delight in company and friendship is sometimes tempered by tedious chores. Occasionally the boredom tumbles into anger:

> *Ian*: I *hate* going shopping in Manchester with my Mum. When I go in C&A I nearly *boil* and I am glad when we come out. I *hate* going in material shops because my Mum spends *hours* looking at material.

Children also contend with changes to familiar places. The Broadoak children had recently lost the only formal play equipment in the area, removed by the council because it had been damaged.

> *Diane*. I think the council ... should give the children somewhere nice and safe to play. I am unhappy and feel right downhearted the way the field has been vandalized and taken down and *not replaced*.

The feeling tone of childhood

Children write of challenging their environments and exploiting them. They write about spaces they invade. In Worsley they scrambled over the heaps of coal in a merchant's yard, tumbled on the bars designed to stop horses getting on to the golf links, and dared each other to tease the local farmer. The golf links were a place of great conflict when the children's play interrupted the adults' games.

> *Rebecca.* After school nearly every day ... the children from our street go on to the golf links. We play very carefully there, because if we run on the green or go near golfers you could get hurt by a golf ball. Some golfers do not mind us playing on the rough, but others do.

> *Jane.* I love going to the golf course. The golf course is very quiet, with beautiful views of grass, and bumps with sand. But when I get told off I feel ashamed and wish I'd never gone.

Children also demonstrate in their writing an inclination to gravitate towards the left-over patches, the forgotten pieces of land which can be unsavoury or even dangerous. One or two boys delighted in recounting rat-hunting trips down the dingle. Others were more circumspect:

> *Peter.* A dingle is dirty and full of germs but it is good for making rope swings on the trees that hang over the brook.

THE QUALITY OF EVERYDAY LIFE

The essays reveal the commonplace aspects of most children's lives, but they also signal the episodes of energy and vigour which characterize middle childhood.

> *Jim.* After I've been scrambling I go home, put my bike away, and go and call for all my mates, and go and have a good game of football on the field. So we all put our kits on and go on the field for a game of football. Of course my hobby is playing football and on our field we have a proper pitch and proper goals so we play football for about two hours.

Jim's writing of his very active life is typical certainly of the older boys, who are allowed a great deal of freedom to roam and indulge in a variety of activities.

Another boy of eleven demonstrated in his writing a contrasting aspect of childhood experience, that of withdrawal and absorbed concern with a particular part of the natural world.

> Another of my favourite places is the first pond on the golf links. It is quite big and has fish in it in the breeding season. In February or late March frogs and newts come to mate and spawn. Also you sometimes get coots nesting there.

THE SIGNIFICANCE OF EVERYDAY EXPERIENCE

These are contrasting images of two boys out of doors, yet they both demonstrate an attachment to, indeed a need for, particular places. A pronounced quality of the children's writing is the expression of their very personal responses to aspects of the landscape. Although humdrum for the most part, often disjointed and constrained, their writing about everyday experiences shows a certain quality of response. Children seek solace in everyday places or delight in the ordinary. In a second piece of writing John tells us:

> I sometimes like sitting under my favourite tree on the golf links when I'm bored or depressed or there's something wrong.

Stories of dens or hideaways are common throughout the essays, from boys and girls in all the age groups.

> *Joanne.* I like to play in my back garden ... behind my shed I keep my secrets when I want to be alone.

Escape holes exist all over the place. 'A secret place is under the tank at the back. You have to have a permit to get in' (*Jeff*). The local library provided sanctuary for one boy from the demands of home life.

But privacy has to be guarded:

> I like our den. My feelings of it are dark, smelly, but good. It is well made and waterproof. It is brick-proof but not Simpson-proof, who often bashes it in.

Simpson was another boy at school. Such pranks by fellow pupils were common, and generally taken in good part. More sinister are the accounts of attacks by older adolescents who roam the open spaces during the evenings.

Running through the descriptions of dens is a sense of the backwoods. One boy had constructed a series of traps to mark out and secure his territory. Eating in a den was a great delight, and some children even slept in garden dens at the height of the summer. One girl extends this craving for independence through introducing an atavistic tone when she writes:

> In the summer I play in a field next door. I make dens in the long grass. The field makes me feel that my parents died when I was young and I have to catch fish and hunt animals.

Children also derive great pleasure from knowing the outdoor haunts and describe them in detail.

> *Dave.* In the Folly Lane-Warren Drive dingle I know all the places to hide in. Behind the bushes, in bushes, up trees, and among tall plants ... it is good for many games.

Learning about these places is itself an important act, as the following quotation suggests:

> *Tom.* I like sitting in the tree in the back of the garden. The tree I climb is full of branches and it is very easy to climb. I never get fed up of playing in the tree because I haven't fully discovered it yet.

Some children use their detailed knowledge to claim ownership and so consolidate this highly personal response to a place.

> *Matthew.* ... and when I get into this tree house I'm there for the thousandth time and I like to think I've climbed that particular tree more times than anyone else in the world.

Much of this knowledge is born of the tactile quality of the environmental experience in childhood. Children tumble and dig, they scramble, and kick, and squeeze into small spaces. They climb and fall, get wet, and get dirty.

> *Jo.* Near this brook there is a tree with a big hole under it and we have lots of fun and it once got flooded and it is exciting enough that I go as often as I can. The cave under the tree is big enough to hold five people.

Another boy's comments capture an exuberant pleasure in the immediacy of contact:

> *Matthew.* Down the path of the golf course on my bike the wind blows in my face. This gives me happiness and on the old station with Ben Simpson over the dunes doing jumps and wheelies this gives me joy.

One feature of the children's writing about contact is their use of exaggerated language. Small mounds become 'hills' and the woods 'forests'. An example from one of the younger children demonstrates this:

> *Mark.* I put on my Wellington boots and go down the brook and jump it. I even walk through the rapids; that feeling is a funny one.

To some extent this reflects the fact that children experience the world at a different scale from adults. A brook looks like a river when you are very small and ordinary potholes feel like 'rapids' when you are wading some distance from the bank.

The children's language also reveals the scope of fantasy in everyday life:

> *Jenny.* I play games in all places, like we play a horror game. We play it in the backings where it is all dark ... My best game is playing Dracula in the backings, where I get my spooks.

Jenny exploits an 'atmosphere' in a place, the dark, narrow back alley between two rows of houses, as a setting for and enhancement of the game. This perception of a particular quality of place is itself a product of children's direct contact. An adult might use the alley as a route way, but the children are there habitually. Adults and children encounter places in different ways, for different reasons, and so the quality of their experiences contrasts deeply.

CONCLUSIONS

But the world is changing. None of the girls in the Broadoak study was allowed to play in the woods alone. The children wrote of warnings about 'kidnappers', 'murderers', and even the proverbial 'Bogeyman'. Ward (1978) and Parr (1972) have both observed and lamented children's loss of freedom as urban areas extend and fields become prairies.

Girls are particularly restricted, and were not allowed to roam and wander like the boys. This restriction is unintentionally reflected in the choice of quotations, for there are far more from boys than from girls.

The world is also more complex. One father I interviewed berated his son, saying how unadventurous he was. 'I used to cycle to Warrington or ride to Crewe on a penny platform ticket.' The cycle track to Warrington disappeared long ago, and jaunts on platform tickets are likely to result in prosecution today.

Despite the restrictions and the complexities, however, this small collection of writings does show that middle childhood still holds a special quality, a 'delighted absorption in our own world, from which in the course of growing up we are distanced and separated so that it and the world become things we observe' (Raymond Williams, 1973).

This short account gives us some evidence of both how and what children write about their environment. Perhaps we might observe that one of the most striking differences between these environmental jottings and writing for geography in school lies in the feeling component. In writing, though for a research project, about their environment the children bring in their feelings freely. This is not generally part of the register of accepted writing in geography lessons. Perhaps we need to ponder more carefully yet the question of acceptable and unacceptable registers and to realize the effect on the fluency, immediacy, detail, and freshness of children's writing. A research project which developed techniques for examining this idea and compared children's writing yielded in different circumstances would be most valuable.

REFERENCES

Jacobs, J. (1961) *The Death and Life of Great American Cities*, London: Penguin.

Parr, A.E. (1972) 'The happy habitat', *Journal of Aesthetic Education* 6:3.

Ward, C. (1978) *The Child in the City*, London: Architectural Press.
Williams, R. (1973) *The Country and the City*, London: Chatto & Windus.

Part III

LANGUAGE AND IDEOLOGY

Chapter Eight

LANGUAGE AND IDEOLOGY IN GEOGRAPHY TEACHING

Rob Gilbert

LANGUAGE AND IDEOLOGY IN GEOGRAPHY TEACHING

This is an odd title, for to speak of language and geography, language and teaching, or language and ideology is to suggest that, somehow, geography, teaching, and ideology are separate from language, and that we can study the relations among them in the way we might study the relationship between, say, vegetation and climate, or place of residence and political allegiance. But of course this is impossible, for we cannot study, think, talk, or write about geography, teaching, or ideology other than in language itself. The centrality of language in constituting our social practices has led regularly to strong interest in the relationship between language and belief, caught most often in the terms 'language and ideology' (Bisset, 1979; Kress and Hodge, 1979).

These developments derive chiefly from a reorientation in the way we think about language. In the past (and for some even now) there was a view that language was transparent, that it was a vehicle or medium for our observations and thoughts, and that well-intentioned and careful scholarship would reveal the true nature of reality through agreement on definitions of terms and how they would be operationalized. The aim was to rid language of its vagueness, to control its connotations, to limit its interpretations, to make it transparent so that the real business of observing and explaining reality could proceed. It may have been the remarkable lack of success of these attempts which revived interest in language, not as a transparent medium which, perfected, would give us an undistorted window on reality, but as a material part of reality, that is, a concrete element of culture which constitutes and does not simply communicate ideas.

Words are things, and ideas are available to understanding only in discourses that are historically produced and always provide the means by which reality is not merely apprehended, but constructed for our understanding. We do not use language to understand the world in the way we might pick up a tool, but rather conduct our affairs in existing discourses which derive from earlier times, constantly in conflict and change, and reflecting the ideologies of their origins. To see language as material is to emphasize its historically produced character and the way that the discourses of everyday life are conducted through it, in contrast to the view that ideas have some essence outside social relations (as in idealist references to 'the spirit of the age' or psychologisms like 'the creative mind').

Ideology as used here refers to representations of human experience and action which constitute our understandings of the world and how it works. Through these understandings we make sense of our own and others' actions, explain motives and consequences, rationalize, predict, and justify events. These representations are the symbol systems through which we engage in the social practices which produce our life experiences. A critical analysis of ideologies will identify how practices constructed by certain symbolic representations serve to restrict the life experiences of people, depriving them of a level of welfare and quality of life they might otherwise enjoy.

For Giddens (1979:6), the critique of ideology addresses 'the capability of dominant groups or classes to make their own sectional interests appear to others as universal ones'. This is reflected in the fact that ideological critique is most important wherever discourse deals with social division. Race, ethnicity, gender, nationality, and class are the major social conflicts around which ideological discourses assemble, and in each case dominant groups attempt to present their own interests as universal ones, especially by denying or disguising contradictions and by suggesting that the existing state of affairs is the only natural one. As a result, the critique of ideology will focus on forms of domination and exploitation in social practices and how such oppression is mediated by symbolic representations, in this case the language of geography teaching.

If we are to avoid sustaining these relationships of domination, we shall need to raise questions about the way explanations gloss over these divisions and inequalities; how they are made to seem inevitable or natural or justified; how ideology can shut out the perspectives of the less powerful. In schools, this means adopting a critical attitude to curricular content, and pointing out the particular ways in which manifestations of ideology operate. In mounting such a critique the concepts at our disposal must themselves be criticized in order to achieve the most comprehensive and accurate understanding of ideology in the curriculum: hence the need for constant scrutiny of the language we use.

Forms and styles of language constrain the ideas which can be developed, for while concepts do not in themselves *determine* beliefs, they can *constrain* them, for not all beliefs can be expressed in a given set of concepts (Keat and Urry, 1975: 128). For instance, a market theory of urban land use can conclude that gentrification either is or is not explicable as the result of individual decisions by agents responding to fluctuations in economic rent. Such a theory cannot, however, conclude that gentrification is generated by the renewal schemes of powerful groups like property developers, government planners, and financial institutions for material gain and political purposes. The discourse of a strict market theory does not include the concepts of political economy which would give rise to that conclusion.

Ideological effects can result from the common tendency for language forms developed in one social context to be transferred to others. Pocock (1971: 21) quotes Burke's observation that English public debate in his time was conducted largely in the language of the common law and real property. A modern parallel is the widespread tendency for the language of economics to supply concepts like exchange, transaction, and market mechanisms to explanatory models in social psychology and political science. Similarly, in geography, varying explanatory discourses develop from different metaphors. Hugh Stretton (1978), in discussing the social theories implicit in approaches to urban planning, identified four main images or models of cities which distinguished the theories: consensual theories of cities as communities; theories viewing cities as battlegrounds for conflict between classes or other groups; cities as market places dominated by the rules of economic competition, consumption, and efficiency; and cities as machinery, emphasizing planning by rational expertise. Each metaphor raises different questions, highlights different problems, and leads to different interpretations.

As geography teachers we have no transparent access to the world. If our practice is to enhance the personal and intellectual resources of the students we teach, we need to take account of the historical specificity of the language we use, how it is produced in situations of conflicting interest and

Gilbert

power, and how particular social relations are constructed through it. Without this critical scrutiny, conventional concern for the accuracy of our measurements or the comprehensiveness of our evidence becomes fatuous. For only by criticizing this language can we hope to control the ideological effect of our teaching.

To illustrate the way ideology is produced in the language of geography teaching, the following discussion cites examples from two sources. First, the language of geography textbooks is shown to be potentially ideological in the way they describe and explain the world (Gilbert, 1984). Second, extracts from interviews with students show how students themselves bring to the classroom certain commonsense usages which further complicate the ideological nature of classroom discourse (Gilbert, 1982). Together these examples illustrate the pervasive presence of ideology in the language of geography teaching.

IDEOLOGY AND THE LANGUAGE OF TEXTBOOKS

Geographical explanation, like most 'scientific' studies of society, has sought to explain social patterns through what Mills (1970) has called 'abstracted determinism', for this form of discourse seems best to afford the detachment and objectivity so prized in scientific practice. Its manifestations are the resort to explanation through abstract forces, as in such concepts as the friction of distance, the ripple effect in changes in urban land use, the factors of production. The consequence of using such terms is that the dynamic of social and economic change is seen to lie in natural forces, so that the influence of decision-making by and competition among influential groups is neglected. The choice of metaphorical terms has the effect of distracting attention from important aspects and practices involved.

Abstract forces are the explanatory tool of geography as a spatial science, but the impact of scientific models is to emphasize how human activity is determined by the tyranny of space. Note the effect of this tendency in the following passage from a geography school book, where modern transport has rendered many earlier settlements unnecessary from the efficiency criterion, though in some cases, for obscure 'social' reasons, alternative provisions may be possible:

> In all three areas which we have considered in this chapter we have noted that too many settlements came into

existence. The landscapes of western countries are relics of the pre-motor age, but they are now changing. Fewer settlements are needed, especially fewer smaller centres. Where low order services cannot be sustained they can be supplied by travelling vans. This enables the hamlet to be preserved for social rather than economic reasons. (Adam and Dunlop, 1976: 75)

Notice how the deterministic language operates here, for unless we ask certain questions the ideological explanation will appear natural, inevitable. The questions we should ask would include: In whose view and on what criterion is the judgement based that 'too many settlements came into existence?' Too many settlements for what? In whose view and on what basis are 'fewer settlements needed'? Why is the economic criterion given primacy and the 'social' criterion seen as a subsidiary consideration? The passive language gives the economic criterion the force of a natural position devoid of conflicting interest, a taken-for-granted assumption about social priorities. If these questions are overlooked we risk ignoring the perspective of groups whose interests might not be served by such priorities.

The connection between passive language and abstract deterministic explanation can be seen again in the following extract from a geography book which presents social change brought about by human action, policy, and economic priorities as if it were a natural process. It is a description of the enclosure movement in eighteenth-century England.

> Gradually, larger farms were formed by joining up some of the open field strips into bigger fields. Fences and hedges were put up to enclose the land. All kinds of new methods were brought in - growing crops in rotation, breeding animals, draining marshy land and using more and more machines in farming. These machines meant that fewer people were needed on the land and so large numbers began to leave the small farming village and move into the growing towns where there were other jobs and more began working at full-time jobs outside farming. (Ayers, 1977:33)

This description is a very selective one, for it fails to mention the division of rural society into farmers and labourers, or the pauperization of country people, or the development of the workhouse to deal with the unemployed. However, the point here is not simply that the ill effects of enclosure are not mentioned, but rather that the way the process is described again makes it seem natural, free of human conflict and power. The passive voice depersonalizes the process, and suggests a functionalist necessity in which 'fewer people were needed' and the rural consolidation was nicely balanced by urban growth. In contrast, the movement from country to town suggests a positive willingness, even choice, on the part of the rural poor, an impression this time created by the active rather than the passive voice. For it needs to be remembered that this was not a natural process or even a piecemeal evolutionary one, but a policy established by powerful groups promoting their own interests, what one historian has called 'a plain enough case of class robbery, played according to fair rules of property and laid down by a parliament of property-owners and lawyers' (Thompson, 1968: 237). Again, by suggesting that enclosure was a natural process in the interests of all, an ideological explanation overlooks conflict, inequality, and the interests of the powerful.

These are not isolated cases, for we find other illustrations of the same denial of intention in explaining human activity, phrasing events in passive terms, where people behave in response to casual forces, rather than as constructive agents. 'Change is an inevitable and essential part of life, and any big change makes problems for the men and women who have adapted their lives to it' (Young 1986:62). 'Working in the steel or chemical industries is a natural thing to do if you live on Teeside' (Young, 1975: 105), while another region can act 'like a magnet, attracting people from all over Britain' (Johnson, 1974: 11).

The earlier description of the enclosure movement described it with an air of inevitability, as if it were the only natural sequence of events. The same could be said of the following textbook passage:

> The ingredients of production, namely raw materials, labour and power, are of primary concern in the location of industry. For extractive and reproductive industries, locations are chosen where natural conditions are favourable. Workers, machinery and power are taken to those places where natural resources are available; and the permanence of the location is determined by the continuance of such conditions as a favourable climate and the abundance and quality of raw materials (Stone *et al.* 1972: 220)

Note here that people became 'labour', and that locations are chosen and workers 'taken' by some unnamed authority. The permanence of the locations is 'determined' by physical factors divorced from human decision or the consideration of human welfare. This tendency to attribute social and economic change to natural forces through the use of abstract and passive language forms is one of the most characteristic aspects of the language of geography.

Further examples illustrate its pervasiveness, as when the effects of agricultural mechanization are said to be alleviated, as 'industry-based towns can absorb the surplus labour force' (Ness, 1971: 7), or when 'The level of unemployment is a good indicator of the economic health of a country' (Ness, 1971: 26). In Cardiff 'The different parts of the city provide homes for different kinds of people so it is possible to divide the city into "social areas"' (Bolwell, 1974: 81). Ironically this dehumanization of people is in contrast to the widespread practice of personifying inanimate objects. Towns have problems, industries move and grow, and regions prosper.

In Kress's excellent study *Linguistic Processes in Sociocultural Practice* (1985) other passages from the discourse of geography and its particular genre in classroom teaching illustrate similar features. In an analysis of a textbook extract on agricultural land use Kress shows how the discussion of the environment of a pastoral region is organized so that different possibilities are weighed up and placed in a particular way, with the result that 'suitability for agriculture' becomes the major concern, establishing certain ways of thinking about nature and the economy, and utility in relation to production. The linguistic forms used privilege particular perspectives, excluding those which might be taken by an environmentalist/ecological position or Aboriginal culture. In Kress's view, 'the text exemplifies a well-known effect of ideology, imposing a prior and systematically organised set of values on nature and on the objects of other cultures ...' (1985: 69).

STUDENTS AND THE LANGUAGE OF THE CLASSROOM

McHoul and Watson (1984) have shown how geography teaching can be seen as the reworking of a commonsense geographical knowledge so that it is relevant to a body of subject knowledge as defined or accepted by the school. Teachers must rely on students' existing resources or quite ordinary knowledge of the 'co-production of relatively more esoteric "subject" knowledge' (p. 28), which is characterized by more formal categories. As shown above, this formal language is abstract and deterministic in form, at least in textbooks. But the translation from commonsense geographical knowledge to the formal categories of 'subject' geography is not achieved easily; most often the translation will be only partial as the meanings of students' commonsense constructions are brought to bear on and play around those given from teacher or text. This raises the interesting question of how ideological constructions from commonsense knowledge play upon the ideological features of textbook geography.

To explore this interaction, passages from the textbook analysis referred to above were presented to students for discussion and interpretation in interview (Gilbert, 1982). Among the interesting results of this procedure was the way in which the passages presented set the terms of the discussion, lending support to the view that the language in which information and understandings are presented to students will provide exemplars of continued discussion which students will take up as models of what formal geographical discussion should be. Take, for instance, the passage quoted earlier on the enclosure movement. When asked to explain the meaning of the passage, one student commented:

> They got all the open field strips and joined them up ... to make larger field strips and they got machinery and all the people had to start leaving because of all the machinery ... (Did the people have to leave?) Yeah, because the machines were taking over. There wasn't enough work for people to do ... (Could anything have been done to provide work?) No, because it cost too much money, to pay everybody when they've got machinery. It's cheaper and it's quicker.

This sequence contains some interesting strategies. At first the student suggests an unnamed 'they' as the movers in the process, suggesting an awareness that this was an intentional change brought about by human action. Even so, those affected had no influence in the matter, as they 'had' to leave. But, when asked for further explanation, the student quickly resorted to attributing the causal influence to machines rather than to the people who owned and used them. Finally, the matter is settled by the invocation of economic efficiency, an understanding brought from the student's own commonsense understandings, since the passage made no mention of cost or speed of production. Another student answered in a similar vein, 'It means that people who used to work on the land as hired hands - their jobs would be taken over by machines. It's quicker to use machines. Because you've got the machine you don't want all those people.'

The point is not that the extracts produced these responses by persuading the pupils of a particular position, but that the selection and construction of language establish a certain way of viewing the events, a perspective which touched off in these pupils additional commonsense explanations based on technological and economic efficiency. Despite the fact that students often revealed a humanitarian sympathy for people's predicament in the passage, it could not overcome the in-built logic of the text. While sentiment might rebel, the logic of the abstract rational framework made inhumanity seem reasonable, the natural thing to do.

Ideological effect is not restricted to the language of textbooks. It can be found in all discourses, and students move through many discursive networks in their daily experiences. When certain perspectives are common across discourses they become dominant and achieve hegemonic status. In this instance the determinism of text explanation and the efficiency criteria of technology and economics in students' commonsense beliefs are connected by their common linguistic features, with clear ideological effect.

IDEOLOGY AND GEOGRAPHY TEACHING

The point of all this is not to suggest that we should divest ourselves of the cloak of ideology, for that is probably impossible. There is, however, much that we can do to strengthen the emancipatory tendencies in our teaching and restrict the oppressive ones. The starting place must be our own practice. Walford (1981) has demonstrated the ideological effect of the very terms in which we discuss geography teaching, claiming that uncritical use of the jargon of justification can be deceptive. For the discourse of rationales for geography teaching is itself a product of past and present conflicts generated by competition among particular social interests (Gilbert, 1987).

Criticism of our own patience can lead to scrutiny of textbooks and resources, and to an awareness of the sources of the commonsense discourses of students' own social understandings. An important means to this end lies in understanding the relationship between language and ideology as outlined here. Ultimately, however, the crucial need is for alternatives to supplant the constraining effects of ideology.

Revised questions, concepts, analyses, and interpretations - in short, a revised language - is a necessary condition if geography teaching is genuinely to serve the humanitarian interest.

REFERENCES

Adam, A., and Dunlop, S. (1976) *Village, Town and City*, London: Heinemann.
Ayers, A. (1977) *Homescapes*, Edinburgh: Oliver & Boyd.
Bisset, N. (1979) *Education, Class Language and Ideology*, London: Routledge & Kegan Paul.
Bolwell, L. (1974) *Wales*, London: Ginn.
Giddens, A. (1979) *Central Problems in Social Theory: Action, Structure and Contradiction in Social Analysis*, London: Hutchinson.
Gilbert, R. (1982) 'Images of Human Nature and Society in the Social Subject: an Analysis of Ideas in the Secondary School Curriculum', unpublished Ph.D. thesis, University of London.
(1984) *The Impotent Image: Reflections of Ideology in the Secondary School Curriculum*, Lewes: Falmer Press.
(1987) 'Curricular contradictions and the social purposes of geography teaching', *New Zealand Journal of Geography*, in press.
Johnson, M. (1974) *Towns*, London: Evans.
Keat, R., and Urry, J. (1975) *Social Theory as Science*, London: Routledge & Kegan Paul.
Kress, G. (1985) *Linguistic Processes in Sociocultural Practice*, Deakin, Vic.: Deakin University Press.
Kress, G., and Hodge, R. (1979) *Language as Ideology*, London: Routledge & Kegan Paul.
McHoul, A., and Watson, R. (1984) 'Two axes for the analysis of "common sense" and "formal" geographical knowledge in classroom talk', *British Journal of Sociology of Education* 5, 3: 281-302.
Mills, C. (1970) *The Sociological Imagination*, Harmondsworth: Penguin.
Ness, T. (1971) *Scotland's People: Where they Live and Work*, London: Heinemann.
Pocock, J. (1971) *Politics, Language and Time: Essays on Political Thought and History*, London: Methuen.
Stone, W., Inch, R., and Lee, D. (1972) *Geographic Fundamentals*, London: Heinemann.
Stretton, H. (1978) *Urban Planning in Rich and Poor Countries*,

Oxford: Oxford University Press.
Thompson, E. (1968) *The Making of the English Working Class*, Harmondsworth: Penguin.
Walford, R. (1981) 'Language, ideologies and geography teaching', in R. Walford (ed.), *Signposts for Geography Teaching*, Harlow: Longman.
Young, E. (1975) *People in Britain*, London: Edward Arnold (1986) *Britain and Ireland: their Changing Patterns of Life and Work*, London: Edward Arnold.

Chapter Nine

THE IDEOLOGY OF GEOGRAPHICAL LANGUAGE

Richard Henley

The role of language in the education process has long been appreciated, if not fully understood. Analyses of language in the educational process have concentrated on language interaction and the role of verbalization.

Very few geographers or educationalists have attempted the type of linguistic or cultural analysis undertaken in the field of literature and drama by Raymond Williams (1975, 1976, 1977). The type of analysis employed by Williams uses a base/superstructure model in which there is a close relation drawn between the control, form, and context of culture (including language and education) and the growth of economic institutions and practices. This does not suppose a simple one-to-one base-and-superstructure relation but a relationship in which culture is seen in a context of social relations and the economy. It is this type of analysis that I believe can be used to try to examine the ideological assumptions that pervade the language used in school geography.

The ideological nature of language has largely been ignored by those concerned with geographical education. The majority of work done so far on the ideology of language in geography has consisted of analyses of texts for racist or sexist subject matter, roles, or assumptions. It is evident that, in terms of race and gender, geography is an ideologically charged discipline. There have been, however, increasing efforts by geography teachers to redress the balance that tilts the subject towards conservatism.

This neglect of language issues by geographers, in terms of philosophy and epistemology at least, has ignored the main streams of thought such as the work done by Giddens (1977) in sociology. This situation has changed during the last ten years as some geographers have sought to confront 'linguistic problems', beginning almost inevitably with Harvey (1973).

Other geographers such as Gregory (1978) and Sayer (1984) have made valuable contributions, but the publication of Gilbert's *The Impotent Image* (1984) has brought the debate on ideology and language to the forefront in school geography.

LANGUAGE AND IDEOLOGY

Geography teachers have come to acknowledge and appreciate the interface between language and ideology in terms of race and gender bias. We accept that the use of certain types of terms and language can portray negative images, reinforce existing social structures and inequalities, and in the light of this we try to modify our practice. This has been recognized at an institutional level by local education authorities who have formulated race and gender policies. I would like to argue, however, that school geography, and the language it employs, reflect much 'deeper' ideological formations.

Language is a social construct and so cannot be viewed as neutral or value-free. Sayer (1984:22) defines language as 'A sharing or transmission of meaning, language is both the medium and product of social interaction and so cannot exist in isolation from society as a whole.' Sayer sees language developing in material processes, most importantly that of labour. The 'preferred' types of language 'inform successful labour' (1984: 23). Marx saw language as directly linking people and work; language 'held together' and 'co-ordinated' people in the work process. All language has a context: the work of geographers, for example, is a conscious activity, usually occurring institutionally, including recording, monitoring, and subsequent reflection, in which language develops as a part of that work process. Seen in this context of work process, it follows that, to a greater or lesser degree, geographical knowledge and language must reflect the structures and context in which they are generated.

Geographical education, language, and knowledge can most fruitfully be conceptualized as part of the hegemony established by the dominant social formation. The concept of hegemony refers to an 'organized assemblage of meanings and practices, the central effective and dominant system of meanings, values and actions' (Apple, 1979:5). Capel (1981) shows that the growth of geography academically and in schools was clearly linked with the need to promote nationalist and colonialist ideologies. As well as having an ideological utility geography has also had a 'mundane usefulness' (Marsden, 1976:12) in which it could fulfil certain needs of industry and commerce. The role of education generally in promoting the integration of the student into society and its class-confirming role is now widely acknowledged (see Miliband, 1973, for example). The role of geographical education in this process today is clearly far more subtle than the jingoistic and overtly ideologically charged geography of the early twentieth century. Geographical education has gone through a whole series of changes since the liberalizing influence of Fairgrieve's 1926 publication *Geography in Schools* (Marsden, 1976: 12-15). The influences bringing about change in geographical education have come from the academic 'trickle down', from teachers themselves, and from wider society, most notably of late in the form of calls for geography to be more firmly articulated to the needs of the economy via more vocationally and technocratically conscious courses.

The nature of the language used by geographers, I argue, reflects the wider social and economic climate and the dominant ideological formations. The notion that geography is value-free or neutral has been widely criticized and shown to be demonstrably false (see, for example, Harvey, 1973; Gregory, 1978; Johnston, 1983; Dunford and Perrons, 1983; Sayer, 1984). It may be that the attainment of a value-free and neutral methodology and language for geography is impossible, but as Gregory (1978: 63) argues, 'Science ... is obligated to be so self-critical if it is to distinguish itself from ideology which I will represent as unexamined discourse ... This demands nothing less than an examination of language itself.'

THE IDEOLOGY OF GEOGRAPHICAL LANGUAGE

There has been a great shift in the meaning of certain 'key words' in the language over the years. Language and meaning vary historically according to both context and user. As Sayer (1984: 94) explains, '"Urbanization" and "Industrialization" mean radically different things at different times and places; for example capitalist and pre-capitalist cities have only the most superficial (and the most asocial) of similarities.' Geographical vocabulary is both subjective and ideological in that it is a social creation: meaning develops within language in relation to social change and social relationships. The use of a term such as 'working-class dwelling' was once widespread but now it has virtually disappeared in favour of terms such as 'inner city'. As a term 'working-class dwelling' has a clear and highly specific meaning. It identifies a specific social grouping and it is loaded with assumptions about the condition and nature of both dwelling and inhabitant. As Gray and Duncan (1978:298) suggest, the meaning of 'working-class dwelling' was too clear, and so government adopted the alternative term 'inner city'. 'Inner city' is a much blander term; instead of identifying certain groups of people it identifies certain areas. The subject and language are both depoliticized and dehumanized in that the area, not the people, is seen as suffering from deprivation.

Foucault (1980) also argues that geographical knowledge and language must be viewed historically in the context, within what he terms an 'archaeology of knowledge'; that is to say, Foucault contends that geography generates few of its own concepts but that they are eclectically collected and absorbed from other disciplines. He goes on to say that geographical knowledge and language are inextricably linked with the exercise of power. 'Region', one of Foucault's examples, was originally a military definition of area derived from *regere*, meaning 'to command'. For Foucault language carries certain 'messages' that can be understood only by isolating language concepts from the groups, structures, or classes that imprison them.

POSITIVISM AND GEOGRAPHICAL LANGUAGE

Perhaps the biggest influence operating on the language of school geography today is the legacy of quantification and positivism; this is clearly apparent in both textbooks and examination syllabuses. The scientism of such an approach is evident. Social relations are treated as objects and variables, not as phenomena in their own right. Bradford and Kent's widely used textbook adopts a pseudo-scientific vocabulary and takes a 'scientific approach' (1977:2). The form of the language and the approach abdicate any notions towards political or social responsibility. Their review of urban geography (pp. 70-85) relies almost totally on metaphors derived from plant ecology. Similarly their review of the gravity model developed from Newtonian physics uses terms such as 'masses', 'friction of distance', 'distance decay function' and 'distance exponent'. Migration as a process is totally divorced from its social, political, and economic contexts. Geography teachers use this form of language constantly. Terms derived from the natural sciences, such as 'boom', 'slump', 'depression', 'trough', 'competition', 'invasion', 'succession', and the like, are all in common use.

Passive phrases such as 'restructuring' which hide very real changes in social relations (such as unemployment, the increased use of 'twilight' shift work, 'union-bashing', and 'deskilling') add a veneer of neutrality and detachment. The language used in geography, although appearing 'disinterested', is in fact infused with ideology. The scientific language employed by geographers seeks to reduce complex valuative issues to matters of technical competence.

Gilbert (1984:47-52) holds that the language and metaphor used in geography hold a great variety of meanings. Society is discussed in metaphorical terms, society is seen as a machine, an organism, a game, or a system. The metaphors are drawn from society and the economy, and the society as a system, currently in vogue, is drawn from the applications of cybernetics in large business organizations.

The language of both science and systems theory seems to offer a more 'powerful' and 'rigorous' mode of description to geographers. The ideology transmitted in such language is one of 'efficiency' and 'control' and of assumed logic in such social and economic systems. Above all, such language dehumanizes and depoliticizes social processes and leads to a general 'flattening of reality'. In such analyses 'natural laws' are seen to operate, and there can be no human interpretation of power and justice.

Johnston (1984) and Lee (1984) have provided a critique of the positivist legacy in A-level geography syllabuses. They argue that the encouragement to see social processes in terms of systems, process, organization, and interrelationships is a restrictive approach that frequently does not go beyond metaphor and analogy. Students are discouraged from seeing social process in any regional, historical, or cultural context. Relations are viewed asocially and ahistorically, and 'problems' are frequently conceptualized in terms of 'dysfunctions' in the system. The language generated by, and informing, the positivist methodology does not encourage students to ask important questions about power and justice. In a critique of the Geography 16-19 project Sayer (1986:89) comments that,

> We can study matters ... such as the influence of government on industrial location but not how capitalist industry itself reproduces a class of people who can lose their livelihood if profits fail and who are denied a say in investment and location decisions.

Students are exposed to, and asked to use, 'indifferent' language, that is, language that is substituted for concrete terms that describe and explain how society works. Geographers seem to have achieved the ultimate in 'indifferent' language in their construction of elaborate mathematical models that, were it possible, would never face problems of meaning and concept.

It seems to be the case that an increasing number of geography teachers find the positivist and quantitative legacy difficult to justify methodologically and in terms of language. The ideology enmeshed in such an approach is conservative and technocratic, and legitimizes the existing social formation. If the ideology is not as blatantly apparent as the jingoism of early twentieth-century school geography many find it far too pervasive and restricting.

LANGUAGE AND HUMANISTIC GEOGRAPHY

Humanistic methodologies have become increasingly important in social science teaching. Sarup (1983:147) sees this partly as a response to the 'anti-humanist severity' of structuralist and positivist methodologies that deny the 'individualistic, the romantic and the subjective'. Humanistic geography has been seen as an important methodological development in addressing problems of human-environment relations and 'our thoughts and feelings ... that result from them' (Fien, 1983:43) or, as Slater (1982) described it, 'a geography of personal response'. Humanistic geography, claim its supporters, is pupil-centred in its concentration on private geographies and environmental experiences. The titles of textbooks such as Beddis's (1982) 'A Sense of Place' series, the use of role-playing and simulation, and the concern with the aspects of perception, consciousness, and experience show that humanistic geography has a growing number of adherents among schoolteachers.

Part of the attraction of humanistic geography is the antidote it provides to the sterility of positivist geography. Humanistic geography has epistemological claims to being empirical (being based on observation), systematic (being concerned with the organization of experience), and rigorous (in that procedures are subjected to a reflective critique). Academic geographers such as Tuan and Relph have done much to popularize the approach, and Gregory (1978) has forcefully put forward a methodology locating phenomenology within a critical theory that acknowledges the profoundly ideological nature of knowledge. Humanistic geography, for all its strengths, still fails to address many of the linguistic problems generated by positivist methodologies. Rose (1981) holds that humanistic geography should have a major concern with language and text. The idea of text interpretation and the concern with language can be seen in the 'interpretative' geography of Lynch, Tuan, Relph, and Lowenthal (Rose, 1981:117-24).

There has been criticism of humanistic and behavioural geography for resorting to what Reiser (1973:205) terms 'Psychologism', that is, the examination of social and historical processes in terms of individual psychological processes. Much humanistic geography seems to be based on the ability to reconstruct experience. In Beddis's (1982) books a large number of exercises ask students to examine or write about their feelings (Book 2:23) or to reconstruct the experience of others (Book 2:93). Similarly the Geography 16-19 project has the 'clarification of personal attitudes and values' as an aim. How is it to be achieved? The emphasis seems to be on the reconstruction of experience and the examination of individual psychological states, often through role play. Too little attention seems to be given in humanistic methodologies to 'value and attitudes' in terms of ideologies produced by wider social forces and structures.

It is important to keep in mind, when considering humanistic geography, Habermas's contention that experience is formed by language (Held, 1980:308) and demands a knowledge of the social context of language and subject if experience is to be understood. The stress on 'experience' and 'consciousness' in Relph's methodology makes it imperative that 'we reflect on our own consciousness and explore the various manifestations of them in our own experience' (1981:101). This assertion seems to beg the question 'Just how are we to make sense of this reflection on our own consciousness and experience?' The notion of *Verstehen* or 'understanding' that is central to the phenomenological/humanistic methodology is essentially based on an ability to reconstruct experience. This seems like a simple reductionism in so far as experience is seen 'organically' and not in context of being a social construct that is mediated by language (Held, 1980:308).

Without a critical analysis of language, humanistic geography is liable to be reduced to descriptivism and the uncritical acceptance of consensus. Without an analysis of language 'nothing can be said about the truth content or possible deception (ideology) expressed by the subject' (Held, 1989:310).

All this may seem far removed from the classroom

geography teacher but there is a need when using humanistic methodologies and teaching strategies to be actively concerned with language and ideology. Pupil tasks concentrating on a 'sense of place' associated with memory, feelings, meanings, and the re-experiencing of personal and non-personal constraints have to be viewed critically by teachers. Humanistic approaches have a clear advantage in developing the student's affective domain, in promoting empathy, and in developing environmental consciousness. A central concern, however, for geography teachers is just how far such strategies go in developing critical thought in students. Without an analysis of language this methodology can easily be infused with ideology, and teachers should be wary of the voluntaristic view of society this approach seems to project. There needs to be an acknowledgement that experience is mediated by language.

Humanistic approaches in the classroom can develop into 'idiosyncratic geography' that never develops beyond the individual's perception and experience. There must be an awareness on the teacher's part that 'ordinary language is intertwined with practice' (Held, 1980:309), that language equally conceals as much as it reveals, and that it is as much an instrument of control as of communication. Unless teachers come to terms with the ideas that language and methodology are both profoundly ideological and can be, and are, used for systematic distortion (see Kress and Hodge, 1979:22, for an account of how newspapers use language to distort reality for ideological ends), then humanistic geography may not develop beyond crude psychologism and descriptivism.

CONCLUSION

Gregory, one of Britain's foremost academic geographers, said that the pre-eminence given to method over explanation has reduced school geography to a 'narrowly technical education: to a set of mechanical exercises' (1981:142). If geography teachers are going to break out of the epistemological straitjacket imposed by logical positivism's commitment to empiricism, then they are going to have to make a 'theoretical effort' (1983:42). Teachers may regard questions of language and ideology as irrelevant, 'too difficult', and as making 'enormous demands' (Gregory, 1981:142) on their time and effort. This may be the case, but, if geography teachers wish to be taken seriously on an academic and professional level, then the 'difficult questions' have to be confronted. Geography has to be subjected to a continued and committed critique. Central to this critique must be a concern for language - to quote Gregory yet again, 'Science ... is obliged to be so self-critical if it is to distinguish itself from ideology, which I will represent as unexamined discourse ... This demands nothing less than an examination of language itself' (1978:63).

At the moment school geography presents an uncritical, passive, depoliticized, and dehumanized view of society. Geography teachers need to adopt methodologies that embody linguistic self-criticism. Students need to be taught to look at society critically and to view it as a structure that they can participate in and change. This does not imply a search for 'methodological purity' on the part of the teacher, but an active concern for the ideological assumptions contained in geographical method and language. Without an examination of discourse school geography will not be able to allow students to participate in or transform their own society and environment.

REFERENCES

Apple, M.W. (1979) *Ideology and Curriculum*, London: Routledge & Kegan Paul.

Beddis, R. (1982) *A Sense of Place*, Books 1-3, Oxford: Oxford University Press.

Bradford, M.G. and Kent, W.A. (1977) *Human Geography: Theories and their Applications*, Oxford: Oxford University Press.

Capel, H. (1981) 'Institutionalization of geography and strategies of change', in *Geography, Ideology and Social Concern*, ed. D.R. Stoddart, Oxford: Blackwell.

Dunford, M., and Perrons, D. (1983) *The Arena of Capital*, London: Macmillan

Fien, J. (1983) 'Humanistic geography', in *Geographical Education: Reflection and Action*, ed. J. Huckle, Oxford: Oxford University Press.

Foucault, M. (1980) *Power/Knowledge*, Falmer: Harvester.

Giddens, A. (1977) *New Rules of Sociological Method*, London: Hutchinson.

Gilbert, R. (1984) *The Impotent Image*, Lewes: Falmer Press.

Gray, F., and Duncan, S.S. (1978) 'Etymology, mystification and urban geography', *Area* 10, 4: 297-9.

Gregory, D. (1978) *Ideology, Science and Human Geography*, London: Hutchinson.

(1981) 'Towards a human geography', in *Signposts for*

Geography Teaching, ed. R. Walford, Harlow: Longman.
(1983) 'Ideology and human geography', in *Geography and Education for a Multicultural Society*, London: Association for Curriculum Development in Geography.
Harvey, D. (1973) *Social Justice and the City*, London: Edward Arnold.
Held, D. (1980) *Introduction to Critical Theory*, London: Hutchinson.
Johnston, R.J. (1983) *Philosophy and Human Geography*, London: Edward Arnold.
(1984) 'The world is our oyster' *Transactions, Institute of British Geographers*, New Series 9, 443-59.
Kress, G., and Hodge, R. (1979) *Language as Ideology*, London: Routledge & Kegan Paul.
Lee, R. (1984) 'Process and region in the 'A' level syllabus', *Geography*, 69, 2: 97-106.
Marsden, W. (1976) *Evaluating the Geography Curriculum*, Edinburgh: Oliver & Boyd.
Miliband, R. (1973) *The State in Capitalist Society*, London: Quartet Books.
Reiser, R. (1977) 'The territorial illusion and behavioural sink: critical notes on behavioural geography', in *Radical Geography*, ed. R. Peet, London: Methuen.
Relph, E.L. (1981) 'Phenomenology', in *Themes in Geographic Thought*, ed. M.E. Harvey and B.P. Holly, London: Croom Helm.
Rose, C. (1981) 'Wilhelm Dilthey's philosophy of historical understanding: a neglected heritage to contemporary humanistic geography' in *Geography, Ideology and Social Concern*, ed. D. Stoddart, Oxford: Blackwell.
Sarup, M. (1983) *Marxism/Structuralism/Education*, Lewes: Falmer Press.
Sayer, A. (1984) *Method in Social Science: a Realist Approach*, London: Hutchinson.
(1986) 'Systematic mystification: the 16-19 geography project', *Contemporary Issues in Geography and Education* 2, 2:86-93.
Slater, F.A. (1982) *Learning through Geography*, London: Heinemann.
Williams, R. (1975) *The Country and the City*, London: Paladin.
(1976) *Keywords*, London: Fontana.
(1977) *Marxism and Literature*, Oxford: Oxford University Press.

Part IV

LANGUAGE AND REFLECTION

Chapter Ten

'A TEACHER, AN ADULT OR A FRIEND?'

Anthony L. Ghaye

The scene is a fairly typical geography classroom in a four-form-entry urban Roman Catholic comprehensive school for children of 11-18 years of age. It is located in an area of Victorian villas and houses built in the inter-war period. Children of Spanish, Irish, Italian, British, and other nationalities attend the school. Its catchment area is a large one, attracting children from a variety of environments such as the inner city, local small towns, and villages.

It is a bright, sunny Monday morning in the summer term. I enter room 2 just before the seventy-minute lesson is due to start, cheerful, expectant, and looking forward to my class of 11-12 year olds. Under one arm my visual aid, an old roll of patterned wallpaper with a large diagram on the reverse side showing the school in relation to a place in northern Queensland called Ravenswood. Under the other, a small portable projector, complete with an extension cable which I've wedged up under my armpit.

In my right hand my briefcase, not full of old committee papers and college progress review documents but crammed with photographs to help convey the notion of a gold rush and hand-outs for all the children in the form of four pictures, in sequence, that reflect the changing fortunes of Ravenswood through time. In the other hand I have a carrier bag. Inside is a battery-powered cassette recorder - too many plugs and cables on the floor is not good news with this group. My tape is ready to turn and at the right place, I hope. Also in the bag is a copy of my lesson plan - carefully thought through, I think - a mug, jar of coffee, some dried milk, and my potato.

I feel rather like a salesman with all this paraphernalia, a little like an ambassador (I'm a college of higher education lecturer back in the classroom again and feeling very much on show), and certainly like an anthropologist as I endeavour to undertake a piece of research that requires me to get inside

175

this culture and explore how the children are making sense of their geographical experiences.

The bell goes, but there is no sign of the children. To fill the expanding silence I reach down into the carrier bag and begin to carve my initials on my potato. (Monday is potato day for staff. The potatoes go in the oven at break time and are devoured at lunch.) No sooner done than the children enter the classroom, cheerful as usual. They always seem to be engaged in such earnest conversation. But here I am, 'tutor-anthropologist', slipping into my professional participant-observer role again. As usual they are never short of something to say as they pass my desk. 'You've had your hair cut, sir.' 'Sorry we're late; assembly dragged on for ever.' 'Have you had a nice weekend, sir?' 'Watch him today, sir, he's in a bad mood. Got out of bed the wrong side, if you tell me.' As they glide past I run through my much practised looks which convey a sense of amusement, interest, slight embarrassment, and a measure of surprise.

As they settle in their places I begin my well used opening move. It is a good exemplar of what Mehan (1979) would call a co-occurrence relationship. 'Good morning, everyone. Good to be with you again.' 'Good morning, sir. Wish we could say the same.' So the lesson begins. First I read a short story, then turn off the lights to play some music from my record of the sound track from the film *Paint your Wagon*. In no time at all we are up on our magic carpet, travelling through time and across space to northern Queensland in the 1860s.

After the lesson the children are reminded of the procedures we have negotiated for writing up and sharing our thoughts and feelings about the lesson, through the medium of their individual diaries. Some of the accounts of the lesson are given here. Taken together, the children are set something of an agenda of issues about teacher-pupil dialogue through diaries that I want to explore in this chapter.

> *Emma.* The lesson made me think about being a banker. I was a banker and I had the biggest piece of gold you could get.
>
> *Paul.* I think the lesson was easy, because it was just like writing a story about a gold rush. I was sad for the people who put up the town, because they had built the town for all the people who were coming to live in the town because of the gold and when the gold worked out, they left, went away and let the town rot away. I was surprised because when you were saying that we were about to go on an amazing journey, on a

magic carpet, I expected to go to the jungle or 'land or paradise', not to a gold town.

Helen. I found describing a day in the life of a gold digger really hard because the digger would have quite an exciting life. So I did quite a few days. What I really want to know is who chose the songs on the tape?

Michael. I found it hard writing about one of those characters as if it were me. I think this was because I hadn't done anything like it before.

Jane. In today's lesson the thing I found easy was imagining things and then writing them on paper because at my drama school we, meaning I, have to sit on the floor and imagine things, I won't go into all we have to do. And then we write about them so I found that quite easy. The hard thing was, when I opened my eyes after I had been thinking about the house which was different colours, you showed us some slides and my image of the house was quite different to what the slides said and also I hadn't imagined a dirt track with lots of trees around and black men with white things sticking through their noses, so that is what I found hard in the lesson.

Sara. The lesson made me think about the way that how hard Tom Aitken must have worked to find gold. It must have taken a lot of thought and skill to draw the diagrams and maps of the mine and the site because I don't think that the education would have been much good. I also think that the men who were probably having a drink in the bar just slammed their glasses down on to the table, raced off to get their horses or donkeys, saddled up, said a quick goodbye to the family and then go galloping off into the hills without saying a word of thanks or praise to poor Tom who found the place. Sometimes my dad rushes off like this. The phone rings and he's gone. Mum doesn't know when he is coming home and gets a bit cross with him sometimes. All this leads me on to what I think about what I would do and how I would feel after a hard day's digging and finding nothing. I imagine it would have been quite sickening watching all the other men loading on to their horses their bags of gold. I should think all the women were spoilt with all those rich men around. I really did enjoy the lesson. I liked the bit when we were imagining things because for once in any lesson we were able to relax. Usually we're always on the go.

John. My views on this lesson were that this lesson was interesting to start but it got more and more boring to the end. I enjoyed the film and the sort of story to it. But what spoilt the lesson was that we had to do a piece of writing at the end of it. It spoilt it because writing about a banker or a miner would not get us anywhere in geography. Maybe in English or history, but I don't see what gold has to do with geography. So could you please explain to me what gold has to do with geography?

Chris. I would tell my friends about today's lesson that when we had the lights off and the blinds down, and it was dark and creepy, there were some boys making scary noises to try to frighten the class. They didn't frighten me, though.

So what sorts of professional issues do these accounts suggest have been, or need to be, confronted, written about, and resolved? What kinds of insights, understandings, and sensitivities about the teaching-learning process do they convey? How far do they reflect the real experience of school?

ISSUE 1. FROM TEACHER-CENTRIC TO PUPIL-CENTRIC PERSPECTIVES

One of the noticeable changes in recent years, in research that has focused upon the ways pupils learn, has been the shift from observations of the teaching-learning process that exhibit a bias towards the teacher's perspective on things to accounts where pupils themselves are asked to describe and interpret their experiences of schooling (Entwistle, 1987). Implicit in this movement are the ideas that children are capable of reflection upon action and that their reflections are worth considering, as useful feedback to teachers about their lessons. It is also considered that introspection is a worthy activity for children in its own right. This latter point challenges the prevailing tendency to judge the value of pupil accounts from the point of view of those who read, rather than of those who own, them. I wish, in addition, to suggest that the pedagogic model currently held by some teachers leads them to underestimate, or even disregard, the capacity of young children to engage in systematic, reflective self-study. An additional complication is the structure and organization of those schools which attribute very low power and status to the child and pupil roles (Calvert, 1975).

Reflection upon action is not just a process; it is also a frame of mind. It requires certain kinds of behaviour and thinking on the part of the children and their teachers that is conducive to exploring classroom interactions and experiences and to being open and sensitive to the problematic nature of the meaning of various social episodes. I have argued and illustrated elsewhere (Ghaye, 1986a) that classroom action has a certain purpose and meaning. It has meaning for the teacher and for the child, and the two are not always congruent. Pupil accounts in the form of diaries can point up the enormous differences that sometimes exist between the teacher perspective and the child's-eye view of the same situation.

What other organization, system or service would deliberately ignore customer opinion? How myopic and insecure we teachers can be. We say we want to meet our children's needs, yet we often decide what those needs are. Then we blame the students for not always responding to our provision. (Brandes and Ginnis, 1986:159)

ISSUE 2. MULTIPLE REALITIES

To generate knowledge and develop understanding we need to put a construction on reality. But what passes for reality? There are many realities. The pupils' accounts of the same lesson, presented above, tend to support this view. They represent different interpretative themes that have been used by the children to impose an amount of order on what is, for both child and teacher, an inherently ambiguous situation (Ghaye, 1986b). For example, in the accounts of the Ravenswood exercise, Paul found the lesson easy, Michael found it hard. Michael has difficulties empathizing and imagining, while Jane found it easy and suggests why. Rachel, on the other hand, thought the lesson 'very good indeed', whereas John thought it got 'more and more boring'. Sara wrote a lot about the substantive content of the lesson and related aspects of it to her own family life. Chris, however, preferred to reflect upon the behaviour of some of his peers.

Insider accounts in the form of pupil diaries can be used to develop a shared model of reality. They reflect a commitment, on the part of the participants, to disclosing usually unknown and unacknowledged realities. They represent idiosyncratic configurations of meaning and may therefore be regarded as unique evidence that helps to supplement the very slender slices of reality that are often portrayed by some of the more conventional formative and summative evaluation practices. But how far does this kind of evidence really get us closer to what experiencing school means? Just how authentic are these pupil accounts of their geography lessons? How can we be sure? How truthful are they? How do we know that pupils, in an effort to be obliging or uncooperative, do not prefabricate an account, do not disguise their real thoughts and feelings, prefer to write safe accounts rather than personal ones? Of course we can never be absolutely sure, as all accounts of this kind are uncertain in that children may be genuinely mistaken about what they think and feel. For other children their thoughts and feelings about geography lessons may always remain inaccessible, no matter how potent the stimulus encouraging their recall.

We cannot escape the fact that a child's diary is a highly selective account. It is not free from bias or distortion. Therefore an important shared responsibility is for the participants to discover the most appropriate way to interpret each account, given the context in which it occurred and which shaped it. A phenomenological teacher posture is, in my view, a prerequisite if we are ever to find a way into the reality of the children's geographical experiences and to understand the subtle textures of meaning which constitute that reality.

ISSUE 3. ESTABLISHING AND MAINTAINING A DIALOGUE THROUGH DIARIES

I have found that authenticity is most closely linked with two things. The first concerns the kinds of relationship that exist between the teacher and each individual child. The second is to do with how each account is handled.

Dialogue through diaries requires of the teacher certain personal qualities, such as tact, humour, a thick skin, a willingness to take risks, and a preparedness to be co-learner rather than expert all the time. Above all else, teachers must be willing to share something of themselves with the children through the diary.

It took me quite a while to find the most appropriate pattern of words that conveyed to each individual an invitation to present their 'loaded commentary on the world' (Britton, 1971). With all I tried, but with some I failed. The following extract is one of the success stories. I have presented five consecutive accounts drawn from the autumn term and five more written in the summer term. There is a qualitative difference in the nature of my comments in Maria's diary and in her entries, across the year. At the start I was unhappy with the way my comments appeared to be creating an accountability context in the sense that I was challenging her to account for her thoughts and feelings. In fact in her fifth entry there is a clear indication that such was her feeling also. On reflection we came to regard this as something of a watershed. We were not sharing in the way we discussed. We were not learning from each other. My intention to 'get data' was uppermost in my mind. I was focusing on product and wanted lots of it. This frame of mind was getting in the way and was putting the whole enterprise at risk. We worked our way through this truncated, impoverished, skeletal kind of dialogue. As our confidence grew, during the year, so did our commitment to maintaining and developing that which Maria owned.

Entry 1

Maria. Today's lesson was easy. I noticed that after the first slide all of them had water or something to do with water in them.
Tony. This is a good entry, Maria. Thank you.

Entry 2

Maria. I thought question (d) was quite hard but (c) wasn't too bad. Its easy to think of different kinds of resources but it isn't easy to think about how people make them or where they all come from.
Tony. Good. Clearly you are thinking about what you are writing.

Entry 3

Maria. I think I found today's lesson quite easy. I found finding groups of pictures on different forms of recreation easy once I had a good label for them. I didn't get stuck this week.
Tony. This is an interesting entry. Could you tell me what you do when you get stuck?

Entry 4

> *Maria.* I think that this week's work carried on from last week's because of the subject, leisure.
>
> *Tony.* I'm sorry you didn't manage to answer my last question. Never mind. Can you please try to write a little more?

Entry 5

> *Maria.* I read your entry sir and I couldn't write any more. You said we could write what we wanted. I'm not trying to be funny but that's what you said.
>
> *Tony.* Thank you for reminding me, Maria, of what I said. I don't think my comments in your diary have been as helpful to you as they should have been. I'm trying to encourage you to say more to me through your diary. But I'm not saying much to you as yet. Am I rushing you, pushing you? What do you think? Let's remind ourselves of what we said, our agreement. You will write to me what you can, in the time you've got, when you want to. I will respond to your entry as fully as I can.

Entry 12

> *Maria.* I liked the way you presented this lesson. I enjoyed it, especially the way the game was introduced as being connected with the lesson even if I went wrong but I didn't make the same mistake again. Most teachers in this school, the older ones more than the younger ones, would never mix a game with lessons. They'd think we were playing around and not learning anything. I thoroughly enjoyed it. I'm glad you're my teacher because we might play some more games.
>
> *Tony.* I'm glad you enjoyed 'Droplet's Journey'. I felt it was a fun way of looking at the water cycle. It took me hours to work out and so I am pleased that you liked it. Perhaps Monday's lesson will encourage you to find some other geography games to play?

'A teacher, an adult or a friend?'

Entry 13

Maria. Trish and me looked in some of the other geography books and found quite a few game-type things but they all looked hard. Could we do some of them in the lessons? Changing the subject now. Sir, how did you come to teach at this school? Was it by your own choice or were you sent here from the college, or what? In September when the teachers get changed around will you stay on, because you are a good teacher? All work and no play, but I won't mention any names. I hope you have enjoyed my entry. I'm sorry about the mess. My little sister dribbled on it. Will you please answer my questions here?

Tony. I came here because I wanted to and because I know Mr D., one of your other geography teachers. Next September I will be back at college helping students to become teachers. I'll miss teaching your class. Can you please check the capital of Australia for me?

Entry 14

Maria. Sir, thank you for answering my questions. The capital of Australia is Sydney. At the college are you a student, teacher, or a bit of both? I ask you that because in your last answer you said 'helping teachers to become students'. Please answer after.

Tony. I'm a teacher. I wonder what you are thinking about your work on Australia?

Entry 15

Maria. Thanks for answering my question. In last week's lesson we were talking about the Australian gold rush. Then we had to write a story about a banker or mine worker or a man who found gold. It was quite good. I'm not very good at comprehension but it was different. I hope the exam isn't too difficult, tests are bad enough but exams are even worse.

Tony. What I felt was really good last week was your story. You really did paint a vivid picture of the fortunes of your prospector. I felt quite sad for him when his gold seam was worked out. Try not to worry about the exams. I believe they should be fun to do, so I hope this one will be for you. Are you enjoying doing your diary?

Entry 16

Maria. Sir, thank you for saying my imagination was pretty good, I don't think so. Trust me to think the capital of Australia was Sydney. I looked it up in the atlas but it didn't tell me much. I agree with you that exams should be fun, not a laugh a minute, but not the kind of thing you should have to sweat out and file back through your memory bank to find an answer. Yes, I am enjoying doing my diary, it's like having a private conversation with your teacher without other people knowing.

Sir, what do you think of how the class has worked, their attitude to work, how they think of you? Who do you think has tried the hardest to work? Has everyone worked well, satisfactorily, badly? Is their attitude that they can't wait to get into their work, or that they couldn't care less? Are they always willing to work or does it take them a while to wake up on Monday mornings? And most important of all how do they think of you, as a teacher, an adult or a friend?

Tony. You've asked me some big questions here, Maria. I'm glad that you feel you can ask them. It could take me a long time to write down my answers. Let's sort out a time when we can talk about them. Perhaps Monday lunchtime in the reading room?

But establishing and maintaining a dialogue through diaries was not enjoyable and rewarding for all the children in the class. The following boys' accounts conveyed this message very clearly.

Paul. I hated doing my diary. I told you I couldn't put my ideas down in writing.

Mike. I think diaries are rubbish, stupid and boring.

Simon. I'm sick of them. I didn't really find the point of doing them. They were not of any help to us. They were only a help to you and I don't know what you needed them for.

Dialogue through diaries soon becomes an ideological and political process. Ideological in the sense that pupils need to be convinced of the genuineness of the teacher's motives and his opinions - for example, with regard to inter-personal relationships, appropriate standards of work and behaviour. For a teacher to say that he sees learning as a partnership, and that sharing thoughts and feelings through a diary is an expression of this partnership, is not enough. There has to be an overt behavioural congruence with this value position demonstrated in such a way that children can clearly recognize it.

It is a political process in the sense that it questions the nature of unequal personal relationships. It questions 'orthodox' teacher-pupil roles, who has the responsibility for what, and who has the power. In short, it questions the roots of a teacher's own 'theory of action' (Argyis and Schon, 1978).

The 11-12 year olds in my class developed a real sense of audience through keeping diaries. They based their accounts upon their perception of me and some hypotheses about how I might react. In other words, the accounts reflected a relationship with a particular reader, and not an unknown audience. Not surprisingly, then, from time to time we all had to be alive to the issue of confidentiality, for what does one censor out of a diary? One illustration of this point is presented here:

> *Hazel.* If I was talking to my friends I would say this subject is fun because in most other lessons we really have to work hard. I'm not saying we don't work hard, but it's different because if I compare English with this subject I would say that this subject is the hardest and English was more boring. I reckon it is the way the teacher presents the lesson. I shall describe some of my schoolteachers.
>
> | | | |
> |---|---|---|
> | Mr A. | = | stern, tall, handsome. |
> | Mrs B. | = | bright and happy. |
> | Mrs C. | = | sulky and very rarely happy. |
> | Miss D. | = | stern and miserable. |
> | Mrs E. | = | small, quite stern. |
> | Mrs F. | = | she's got a lot of guts. |
>
> Please don't show these teachers this. Thank you.

There were some teachers in the school who viewed the sharing of experiences through a diary as a 'subversive' activity and were prepared to tell me so. Generally reactions ranged from genuine interest to degrees of horror, scepticism, or incredulity. Reactions were sometimes hard to predict and difficult to handle. The following extract from the notebook of Mr D., the head of geography, an important participant in the whole enterprise and a good friend, articulates his position in relation to this issue.

> I really can't help worrying what most teachers would think about this. My hunch is that it would be frowned upon. Even more so if someone wrote, 'That was a hell of a boring lesson', I mean, it might have been, but you don't want them telling you that. It could create enormous problems for some teachers, to accept or even read the comments pupils make. It might also create problems for the children, who might feel torn between writing truthfully, in their diaries, really speaking their mind, and writing safe responses that won't offend their teachers. In the end I guess it all boils down to the individuals involved.

On reflection I learned that teacher and child must be clear about what it is that is being exchanged. Initially, to help the children with this, I attached twelve questions inside the flap of each diary. They covered a range of issues about the teaching-learning process. Technically the entries that responded to one of my questions were not truly unsolicited accounts. I learned to live with this, however.

Most important was the invitation to write about something that the children felt was of significance to them. During the year more and more children took up the invitation. Second, the children must be clear about how their accounts may be presented. How far can they be free of literary conventions? For some of the children in my class it became a habit to write a few lines and then draw a picture. Third, there needs to be clarity about how accounts can be most appropriately exchanged, and finally why such an activity is being encouraged in this context. What are the rewards? How do the participants get pride, pleasure and a sense of satisfaction from doing it? Who benefits?

ISSUE 4. LEARNING ABOUT ONESELF AND OTHERS

Other researchers have addressed the issue of the purposes and dangers of keeping diaries (Woods, 1986; Hopkins, 1985). But what I wish to do is stress the need to see the keeping of diaries from both the teacher's and the child's perspective. For me the understandings and sensitivities I gained affected both my espoused theory and my theory-in-use (Argyris and Schon,

1974). I soon began to appreciate that if 11-12 year old children were given the opportunity, they were more than capable of self-conscious reflection upon and appraisal of their experiences and my actions. Their perceptions were welcomed. Their accounts provided useful feedback that helped me to modify my teaching. They also affected our relationship. We began to acknowledge the fact that we could be honest, open, and frank with each other. But at least we could share this privately through the diary or more publicly during the lessons, happy in the knowledge that we were doing so within a supportive framework of interpersonal relationships. Some of these perceptive and potent accounts that affected the way I organized and delivered my lessons, and which also had an affect on my self-concept, are given here.

On children working together in groups

Emma. The thing that really annoyed me was that Rachel seemed to think she was the leader, she had to do everything that was easy (like cutting out) while the rest of us had to do all the horrible jobs like sticking the shapes on. (I'm not saying I don't like her though!) Also when you said that someone has to talk about the map, Rachel suddenly didn't want to be the leader. (I would have been but they made me leader last time.)

Nicola. My opinion of today's lesson was that it was very interesting in the way that it was really like solving a problem in real life. It was also nice to work in a group and trying to discuss some ideas. In fact our group spent nearly all the lesson trying to think how to solve the problem given.

Heather. Today's lesson was good fun although we did not really have enough time to finish it. This was mainly because in our last lesson we spent too much time thinking of ideas instead of getting down to it. Also the lesson was different to some of the others and I myself preferred it. I suppose one reason was because we were working in a group. This makes life easier because we can communicate our ideas with the group so we know how our idea is going to turn out. So yes I did like the lesson today and this is what I thought of it.

John. In the lesson today we were in groups which made it slightly harder. The reason being was because we all had different ideas and we had to decide on two things. At least we had lots of ideas. Also when we were working together we got on better because if we didn't know what to do someone else did. The only problem we found was that some people were dashing on and some were still thinking their own things out. Otherwise it was an OK lesson. Don't forget everything gets messy and out of hand in large groups, will you?

On my presentation of the lesson

Colin. I feel you presented the lesson all right but you don't get straight down to the point, e.g. when we did the histogram I was not sure what you meant so I asked whether it was a columns graph or just different lines. I would say draw a histogram which is like a column graph.

Paul. I find it hard when you just told us something and then we have to write about what they have talked about. That is one of the hardest things to do and boring too. I have to remember it and if your memory is no good that day, then you don't get a good mark. I can't remember things very well but I am getting new things in my head every day.

Christine. I don't like the way you presented the lesson because you are silly to ask us to take our bags off the table and get all the stuff we need out of them. Because if we didn't we wouldn't have anything to work with. Another thing I don't like is your writing. I have to ask my friend to read it, she can hardly read what you put, either. For example, last week she said, 'Look how he's wrote interest. It looks like he's wrote intect.' That will be all for this week. Bye.

Joanne. The way I would change the lesson and how you presented it is that you told us to get on with our work, then every five minutes you stop to tell us something. It is good in a way but we don't get much done if you keep talking. You said if we get stuck that you would help us, but you don't. You stop everyone and tell us that we were to do the work ourself. I think it would be better if you told us what to do at the beginning of the lesson and if you were going to help then you should help us.

Clare. What really interests me about the things that teachers do is the way they explain things, whether or not they use any variety in their lessons and whether or not they explain things clearly. If they explain things clearly they are usually much nicer people you can get on with more easily and you get more work done. If they don't explain it clearly they aren't normally such nice people and they expect you to get it right even though they don't explain it to you. I don't like people like that. This does not mean to say that I don't like you, as you are a nice teacher. (You didn't have a chance to explain much today anyway as we were doing all the talking.) But in my opinion if a teacher has a chance to explain something and doesn't and not just once or twice but all the time, they aren't such nice teachers.

ON VIDEO-RECORDING GROUPS OF CHILDREN AT WORK

This was part of a procedure that I developed for the analysis of pupil interactive thoughts. Video feedback was used in a catalytic medium, later in the same day, to help groups of children to re-experience and to explicate the rationale upon which their actions, in their group work, were constructed (Ghaye, 1986a).

> *Robert.* I was sad because our group was not being filmed when last week you said that we would all be filmed. I found it amusing when we could see the people that you were filming on the screen. I also found it amusing when Paul G. kept pulling faces and messing about behind you.
> *Kate.* I think the bit I was angry about was the way Mr D. kept the camera on the one particular table. I'm not saying that I want to be the star attraction of things but I don't think it's fair on the other kids.
> *Clare.* The thing that surprised me in the lesson today was that the camera didn't stay on one table for all the time. The things that annoyed me last time was that you picked on special people as though you didn't care how we progress, not that I want to be asked questions that I can't answer but you did always make a film of these very special children. If I was a teacher of any sort the thing I wouldn't do is have teacher's pets. I don't want to be in this special group because I don't like being teased because everyone will be calling me

> teacher's pet.
> *Russell.* I would love this lesson to carry on another time because I enjoyed being on the TV camera and being able to watch myself afterwards. I would love to be able to be on TV again as I love to watch myself as a star on tele afterwards.

What became clear to me during the year was that the diaries were performing a metacognitive function, as they were helping children to become increasingly aware of the social *milieu* in which their learning was taking place. Dialogue through diaries helped them become more aware of what they were doing and feeling, and of how and why they were acting in a particular way. Diaries, for some, became a way of monitoring their own learning. I see this development of metacognitive skills (Nisbet and Shucksmith, 1986) as one of the important functions of keeping diaries in this context. A repertoire of such skills may also be characteristic of a successful learner (Holt, 1964). This view emerged strongly when, at the end of the year, I asked each child to respond to the request 'Write down the single most important thing that you think you have got out of doing your diaries this year.' The largest single response (29 per cent) was that the diary was an opportunity for self-disclosure, for 'saying what we really think'. But interestingly the second largest response (23 per cent) was 'Reliving classroom experiences through my diary helped me get better at learning.' For example, Emma wrote:

> Many people would say oh that's easy to do, all you have to do is scribble any old thing down, but you don't. You have to think hard about what you are writing. You have to recap over your work. It forces you to remember. It helps you revise, as you can write down what you did in your own way, what you were doing and it explains it.

SOME REFLECTIONS ON A PROCESS

The major intention of this chapter has been to give the reader something of the flavour of the nature and advantages of children and teachers writing about their geographical experiences through the medium of a diary. However, at this point I wish to raise what I regard as four of the potential problems of doing this. First, for the teacher there is a danger that diaries may slide over and become a celebration of self, an exercise in self-indulgence with a high therapeutic value.

The general problem that I am alluding to is that we must be mindful of valuing the accounts the children write only from the point of view of those who read, rather than those who own, them.

Second, diaries can all too easily be seen as a covert way by which teachers assess their children. In this situation dialogue through diaries becomes a kind of game, where children adopt strategies to search out the teacher's understandings, and where they move towards the teacher's system of relevances. Accounts of this kind convey something else and something less than accounts that are shared in a truly 'participative' spirit (Elden, 1981) where the teacher and child both have a commitment to enriching their shared understandings and to maintaining that which is the child's.

Third, without this participative spirit there is a problem that what the teacher claims is reality in the mind of the child is not reality in the same way the child perceives it. Moving towards a more holistic view of classroom interactions means that teachers have to get used to the idea of working, more often, from the child's definition of reality. They also need to question the assumption that they know what this is to begin with.

Finally the diary may be used as a means of offloading all sorts of thoughts and feelings about important incidents in the child's life history, such as incidents with family or friends in the home or out of school hours. This poses an information handling problem for the teacher. In my view the teacher and child should constantly discuss the kinds of account that are mutually beneficial and desirable. But drawing lines of this kind is a delicate issue, a complex professional problem, and something that the participants have to learn.

Inevitably there will be difficulties. But whatever the problems, establishing and maintaining a dialogue through diaries is an opportunity for teacher and child to see familiar classroom events with new meaning. It is an opportunity to probe beneath the surface realities of classroom interactions. It is a chance for all the participants to come to know and appreciate significances that normally remain unknown and uncelebrated. Perhaps most important, this kind of activity provides an opportunity for forming significant friendships. This means more than children having respect for their teachers and teachers empathizing with the children. Significant friendships are based on trust, tolerance, and fairness between teacher and child. Dialogue through diaries can promote these general values, which I believe to be critical and fundamental to any educational enterprise.

REFERENCES

Agyris, C., and Schon, D. (1974) *Theory in Practice*, San Francisco: Jossey-Bass.
—— (1978) *Organisational Learning: a Theory of Action Perspective*, Reading, Mass.: Addison-Wesley.
Brandes, D., and Ginnis, P. (1986) *A Guide to Student-centred Learning*, Oxford: Blackwell.
Britton, J. (1971) 'What's the use', *Educational Review* 23: 205-19.
Calvert, B. (1975) *The Role of the Pupil*, London: Routledge & Kegan Paul.
Entwistle, N. (1978) *Understanding Classroom Learning*, London: Hodder & Stoughton.
Elden, M. (1981) 'Sharing the research work: participative research and its role demands' in P. Reason and J. Rowan (eds.), *Human Inquiry: a Sourcebook of New Paradigm Research*, Chichester: Wiley.
Ghaye, A. (1986a) 'Outer appearances with inner experiences: towards a more holistic view of group work', *Educational Review* 38, 1: 45-56.
—— (1986b) 'Pupil typifications of topic work', *British Educational Research Journal*, 12, 2: 125-35.
Holt, J. (1964) *How Children Fail*, Harmondsworth: Penguin.
Hopkins, D. (1985) *A Teacher's Guide to Classroom Research*, Milton Keynes: Open University Press.
Mehan, H. (1979) *Learning Lessons: Social Organisation in the Classroom*, Cambridge, Mass.: Harvard University Press.
Nisbet, J., and Shucksmith, J. (1986) *Learning Strategies*, London: Routledge & Kegan Paul.
Woods, P. (1986) *Inside Schools: Ethnography in Educational Research*, London: Routledge & Kegan Paul.

ACKNOWLEDGEMENT

I am enormously grateful to the children of the Eastbank School for allowing me to use their accounts once more.

Chapter Eleven

WRITING AS REFLECTION

Margaret Roberts

INTRODUCTION

My own interest in 'learning logs', or diaries, began several years ago when I was reshaping a Geography Method course for PGCE students. One of the assignments for the course asked students to analyse different types of writing by pupils during teaching practice. I particularly wanted students to encourage a reflective type of writing in children. I was interested, through the work of Douglas Barnes, Pat D'Arcy, and Michael Armstrong, in pupils using writing as a means of learning, as an opportunity for sorting out ideas, for exploring uncertainties, and for shaping tentative understanding of new concepts. Most writing in school geography is not of this type. On the contrary, it is usually the end product of learning rather than the means, and pupils have a vested interest in concealing their ignorance and denying their confusions.

When the students all failed to produce any examples of reflective pupil writing I realized that the model the Division of Education was providing at the time was quite a contrary one. Polished essays were expected, carefully argued opinions were welcomed, but nobody expected any written account of the confusions and uncertainties which exist in any learning mind. How could I expect the PGCE students to collect from pupils a type of writing for which there was no model on the course? That was how I came to ask students to write their learning logs.

The learning logs, however, also became a solution to another set of problems which concerned me and which are common to all teachers. I wanted to know what sense students were making of the course. The second time I ran the course I had made changes based on my observations and feelings about the first year. I had several pieces of evidence to use on which to base decisions. (1) I knew how students had reacted and participated within the sessions. (2) Students often make direct comments. (3) Watching students on teaching practice gave clues about their interpretation of the course. (4) At the end of the year the students wrote an evaluation.

Yet all this was not enough. The evidence was partial and haphazard. The clues were stronger for some students than for others. What was really going on in their minds at the start of the course? How did the approach of teaching practice affect the reception of ideas? At what stage of the year was it appropriate to introduce particular concepts? How did a period of teaching practice affect their thinking? I wanted to eavesdrop on their minds. I wanted to understand thoroughly the process they were going through. Evidence on teaching practice, or an evaluation at the end of the year, marked end points in learning. I wanted to understand the process as it took place.

Although it was my interest in written work which originally induced me to ask the students to keep diaries, it has been the way the diaries have given answers to the problems outlined above that has been really exciting. Why had I not used them earlier? Why weren't all teachers using such a valuable device? My own enthusiasm prompted one of my students and two local teachers to use them with sixth-form geography groups. This chapter is an account of the use of learning logs with PGCE students and with these sixth-form groups. The writing in the logs is quite different in each of the four experiments I am describing. This raises questions about the role of the teacher/tutor in influencing what is written.

EXPERIMENT A. LEARNING LOGS ON A PGCE COURSE

Learning logs became a compulsory assignment for all students on the Geography Method course. As with all other assignments there were written instructions in the method handbook:

GEOGRAPHY METHOD ASSIGNMENT 1
Learning log
During Term One and Term Two keep a diary of your thinking about teaching Geography. The diary will be confidential, to be seen only by you and your method tutor. You should make entries in your diary after each Method Session. Entries should be handed in so that tutor

comments and replies can be added. You may reply to these comments in your next entry.

The diary should record your reflections on the session and might include the following:
- comments on ideas which you think useful
- comments on ideas you would like to try out for yourself
- ideas of your own generated by the session
- comments on things which you consider irrelevant or impractical
- details of what you found difficult to understand or do
- fear, hopes, anxieties
- comments on methods used in the session (would you use them yourself?)

The purpose of the log is twofold. Firstly, it should help you to clarify your own thinking, to identify misunderstandings and to reflect upon the course as it takes place. Secondly, the log should form part of a dialogue between tutor and student through which both can learn more about the process of initial teaching training. The hoped-for increase in understanding should make it possible to adjust the course to suit student needs, and to identify individual student needs.

In addition to these written instructions emphasis was given to the idea of the writing being exploratory rather than descriptive and evaluative. The assignment had to be completed as a course requirement but would not be given a pass or fail grade.

The logs were to be in two parts, corresponding with the first two terms of the PGCE course. In term 1 the students had five weeks at the Division of Education before the first period of teaching practice. In term 2 the students returned to the university for a further four weeks before their second period of teaching practice. The learning logs were not continued for term 3.

The students wrote in their diaries after each of the method sessions and handed in all the entries before each teaching practice. Reading them for the first time was a revelation; it was a breakthrough. Both reasons for introducing the logs were amply justified. First, the writing in the logs was valuable to the students themselves and became part of their process of learning. Second, the logs helped me to understand the process of the course.

1 Value to the students

Reconstructing knowledge

In their writing nearly all students related ideas introduced in method sessions to their own experience. In the first term it was to their own schooling that they referred most frequently:

> I remember when I was at school that in some subjects, especially languages, we were assumed by the teachers to know more than we did. Part of this, I think, was that we did not have grammar lessons as much for English and therefore did not know such terms as dative, accusative which the German teachers expect us to know already and therefore never explained to us. So I stayed in a state of total confusion.

In the second term the main point of reference became their experiences during teaching practice:

> The thing I found most difficult was not leading the pupils to the 'right' answer and also thinking that because one person knows the answer that the whole class does.

Their anecdotes, generalizations, and judgements were valuable because they were encouraging the students to ask constantly, 'Do these ideas match my own experience?' The students searched for evidence to accept or refute the new knowledge. Writing about new ideas helped the students to analyse and reconstruct their experiences afresh to give them new meaning. The more links they made and the more questions they asked the more likely they were to make sense of the course. The learning logs promoted this process.

The students also made links with the future. Before the course started they had some sort of picture of what teaching practice might be like. The assumptions behind the picture made some ideas acceptable and others risky:

> I'm not sure, I would use the 'cake' idea ... perhaps I'm just being 'chicken' not wanting to use it but I suspect I will have enough problems in lessons without having to cope with 'It's not fair!'

An interplay was shown in the logs of students adjusting their images of the future to accommodate new ideas, or of adapting or rejecting ideas because of their own images. The logs that were most valuable to the students were those in which there was a constant looking backwards and forwards, showing a continuous attempt to make sense of new knowledge in term of past and imagined experiences.

Clarifying thinking

Most students took the opportunity offered by the logs to write down their ideas on a wide range of issues. Some merely reiterated what they had already said in the sessions, but most continued to develop their arguments. The process of writing made them clarify their views on issues or occasionally helped them to understand the meaning of the sessions.

> As I have been writing this I suddenly realize the value of the exercise ...

Some felt that they had been misunderstood in the session and used the log to state more clearly what they had thought:

> I think I may have given the impression in this session that I am a raving fascist ... I would hope this is not the case.

For those who were reluctant to join in the discussion the log became a voice, a vehicle for joining in the debate, albeit one-sidedly, at a later stage. The logs showed that many of the silent were more actively engaged mentally in the discussion than some of those who always had something to say.

Expressing feelings

The logs also became an outlet for expressing feelings which were often strong and complex. Some anxieties were expected:

> Control of a class during a video programme must be quite difficult.

Other worries were confined to particular individuals:

> A pet fear of mine is to be asked a question I know nothing about.

The logs helped students to identify their anxieties, to reflect on them, and to share them. It was equally valuable for students to write about positive emotions. By writing about excitement, enjoyment, and satisfaction they were reaffirming their commitment to teaching.

Many students have confirmed my analysis of the value of the logs to them. Some became involved in them:

> I have found in the Monday sessions that I have been thinking of things I will want to write down in my log. I have made notes in pencil in the session so that I will not forget what I have to say.

Others wrote explicitly:

> The learning log has been a useful vehicle to think through your own ideas and to clarify what has been learned.

2 Value to the tutor

Understanding individuals

There are many opportunities to get to know students well on a PGCE course. Discussion and group work play a large part in the method sessions, and there are long discussions with students during teaching practice. So it was somewhat surprising to find out how much more the logs revealed about individual students. Many wrote things they would not or did not disclose either publicly in front of other students or privately in conversation. All the ways identified above in which the logs were valuable to students were equally helpful to the tutor. References to past experiences, interpretation of visits to schools, and what students chose to describe of their teaching practice contributed to an understanding of the context of their educational thinking. Their own experiences became a useful common point of reference. Knowing what worried a student made it possible to explore feelings more sensitively, to listen, and to give constructive advice. In the case of the students who contributed infrequently to discussion the logs became the main way of knowing what they were thinking and feeling, and they gave a starting point for a dialogue.

Understanding the process for all students

Although individuals varied in what they thought and in what they felt, there were many common elements in the logs which gave insights into the rhythms of the year.
For example, there was a pattern in the type of references to the past. In term 1, references were almost entirely to pupils' experiences, even when the logs referred to observation during the visits to schools. An understanding of the pupil perspective is, of course, essential for a teacher, but the dominance of this perspective was worrying. Did it mean that during all the activities in the sessions they were imagining themselves more in the pupils' than in the teacher's role? This particular insight led to a change in the programme for term 1 to include some method sessions in schools where students worked with small groups of pupils. The same issues were dealt with but in a situation which forced recognition of the teachers' perspective and the teachers' role.
It became possible to understand more fully from the logs the changes in feelings at different stages of the course. Anxieties built up before first teaching practice, and were more varied than the accepted wisdom that the main worry is about classroom control. The increased anxiety influenced the type of session which was possible just before teaching practice.

> Quite naturally at this stage, all thoughts are on the next week.

Theoretical issues, which could be the focus of heated debate in the early part of each term, were not welcomed. The emphasis had to be on practical, applied work.
Term 2 was remarkably free of anxieties in the logs. Instead there were different emotions to take into account: an impatience to be back in the classroom; an urge to apply what had been learned in one school in the next school; and the need to have full recognition of the fact that the students were now experienced teachers (if only of six weeks' standing). The logs gave access to invaluable inside information and informed judgement about how to structure the course. It was quite clear that the course could not be regarded as a series of interchangeable slots. It was a progression, but with a great leap between terms 1 and 2, and changes in mood as the terms progressed.

Direct evaluation

An obvious way in which the logs were useful was in their evaluation of activities. They gave a clear picture of what students felt was a useful way of spending contact time. It was definitely not doing analysis which could be done at home.

> I did feel, however, that the continued analysis of a series of extracts became rather repetitive and indeed boring after a while.

Discussion, group work, and variety were all approved:

> The pooling of information and group discussion has led to a higher level of personal thought and motivation on my part.

There were some surprises. For instance, in one session students had role-played difficult classroom situations. The session had been fun for everybody and appeared to have been successful. The logs showed that the students <u>had</u> enjoyed it but had not found it useful. The logs explained why.

As the students became more confident, as well as giving criticism they gave advice and suggested alternative activities.

Indirect evaluation

Another way in which the logs judged the course was in the amount of writing after each session. Generally the more the students had participated as a group the more they wrote. Heated discussions always provoked the longest entries. When there was more input, and in theory more to write about, they wrote less. The mental processes had to be fully engaged in the session for the writing to flow. So, paradoxically, there was less response on the topics and issues where the tutor had most to contribute in expertise because there was a tendency to overload the sessions with input. A salutary lesson!

Each year the course has been modified in response to the direct and indirect evaluation. In this way the logs are helping to monitor a constantly evolving course.

EXPERIMENT B. SIXTH-FORM FIELD TRIP

One of the PGCE students who had been writing her own learning log on the method course was keen to use the technique herself as a teacher. An opportunity arose when her teaching practice school invited her to accompany a sixth-form residential field trip to the Lake District. All the students were given a small notebook in which to write their logs and in which they jotted down the initial instructions. One student's notes read as follows:

include:
1   Anything found confusing/difficult/easy/why doing it.
2   Your comments on how fieldwork presented by staff - good/bad points.
3   What you feel you've understood better by actually doing it.
4   How is field trip different from being in the classroom?
    (a) educationally
    (b) socially
5   Attitudes to:
    (a) geography
    (b) school mates
    (c) teachers
    (d) education/school
6   Any general feelings.

The notebook logs were different in several ways from those of the PGCE students. Entries were fairly brief, perhaps because the small page size suggested a limited entry, and perhaps because they were completed late at night after a tiring day in the field and an evening of recording work. No students made any links between their geographical experiences on the field trip and any other experiences. Very few used the log to sort out their ideas or to explore misunderstandings. An exception was this entry:

Chi calculations/degrees of freedom - freedom from what? This is too mathematical for me. I do it without fully understanding it.

This was the only entry in which a student asked a question on what he learnt. Only one used the log to express a geographical opinion:

> A lot more can be done to protect the country and strong laws should be introduced.

In general the students were not using the writing in the logs as a means of learning. They did, however, welcome the chance to give the teacher feedback on what they felt about the course, educationally and socially. Through the logs the teachers learned what individuals found difficult:

> I found difficult the Chi Square exercise because it dealt with so many different figures.

Teachers also learned that students needed to be clear what the aims were:

> I thought the exercise was pointless as we didn't know what it was leading to.

and that new ideas need time:

> I think we are doing too much work too quickly and I have problems absorbing it.

These few extracts cannot give the full flavour of the wide range of topics covered, from opinions on bedtime, feelings about climbing mountains and measuring path erosion, comments on other students and on rule-breaking, etc. The logs helped the teachers to understand what was going on in the sixth-formers' minds during the week. The logs were full of judgements which would help a teacher plan a similar course. From the teachers' point of view the logs were valuable.

Why, though, wasn't the writing more valuable to the students? Why was there no reflective writing? There are four possible reasons. First, the instructions were fairly precise and they were numbered. Students tended to answer each point every evening. No students wrote anything outside the suggested framework. This links with the second reason. The instructions had been framed by the teacher to suit her agenda. It is thus almost inevitable that the logs would be more valuable to the teacher. The students were writing what the teacher wanted to know and not what they wanted to tell the teacher. It seems that vaguer instructions or an indication of the purpose of the logs would be more important in encouraging students to write for their purposes. Third, students are not normally expected in school geography to write reflectively, so they may find it difficult to write in a manner totally opposed to the conditioning they have received at school. Fourth, the kind of writing in the logs might depend on the pedagogy of the sessions. On this field trip it seems that most of the time was spent on task, collecting data, recording and analysing results. There seemed to be little time to stand back and consider what had been learned, what issues and questions it raised, how it linked with other experiences, etc. Perhaps the rationale of the learning logs needs to be paralleled by the rationale of the pedagogy. If students do not and are not expected to ask questions and reflect on their learning during sessions why should they do so in their logs?

In both experiment A and experiment B the response by the teacher to the students was by discussion and after completion of the entries. The main variables in these examples were in the way the instructions were given and in the pedagogy. In the next two examples another influence becomes significant: the way the teacher replies to the logs.

EXPERIMENT C

The third use of learning logs developed from a teacher's interest in evaluating his use of the new Schools Council 16-19 geography syllabus. He had already been using a questionnaire at the end of each unit. The learning log was seen as an extension of this, encouraging a wider range of writing and more detailed feedback.

The first use of the logs started in June with a group who were accustomed to completing the questionnaire. The following written instructions were given:

> At the end of each lesson your teacher will allow you five minutes to note down your thinking about the lesson which you have participated in. You should try to note down your thoughts on the following topics:
> - ideas in the lesson which you think are useful
> - ideas which you would like to try out for yourself
> - ideas of your own which were generated by the lesson
> - comments about things which you considered irrelevant or impractical
> - details of things which you found difficult to understand or do
> - comments on methods used in the lesson
> - anything else about the lesson which you consider important to note

> Put your name at the top of the sheet and hand it to your teacher. We will try to comment on your notes and you may reply to these comments in your next entry.

The writing in the logs was disappointing. It fell into three categories: (1) *descriptive*, a statement of what had happened in the lesson; (2) *evaluative*, judgements of aspects of the lesson; (3) *reflective*, exploration of ideas. Almost all the writing was in categories 1 and 2. The following extract was typical:

> Today's lesson was simply spent going over the decision making exam we did. It was useful and helps you realize areas where you went wrong.

The descriptive statements are of little interest to the teacher, who already knows how the lesson was spent. For students the only purpose of description is to provide a context for any other writing. The evaluative words were 'useful', 'difficult', and 'irrelevant', all taken direct from the instructions. The evaluative comments are providing the same kind of information as in the questionnaire, but in detail for each lesson instead of for a whole six-week unit.

Very few students used the logs to record their own geographical thinking, let alone explore it. Those who did, did so infrequently. Some got quite close, but then stopped:

> I thought this lesson was quite interesting as it made me think about the impact of oil on the environment.

Well, what did it make him think? The log does not say. A small number of students used the log to admit ignorance:

> I didn't understand the idea of the oil being mixed in with other substances and how the oil got into the rock in the first place.

Why were the logs not as reflective and illuminating as had been hoped? Perhaps the group's expectations of what to write had been influenced by the previous questionnaires. Also the instructions encouraged evaluative comments more than reflection. Although feedback had been intended, the teacher had in fact not written comments after the entries.

If the group's expectations were not conditioned by a questionnaire, if the instructions were changed, and if the teacher added comments, would the type of writing change?

Writing as reflection

He decided to try again with a new sixth-form group. The instructions were changed to focus more on what was in the student's mind. The teacher decided to write a comment after every entry. After six weeks the logs were evaluated. To what extent had the new instructions and the comments influenced the writing?

There were some changes. Most pupils still described what had taken place in the lesson, but some were picking out details which were apparently significant for them:

> We had to choose a factor which we thought best illustrated urban deprivation. I chose overcrowding but we found it difficult to make the initial choice and we had a slight argument to decide this. A lot of people used the factor of not having a bath. This shows that people have different opinions.

This is different from a general summary in that it is beginning to show glimpses of the thinking activities going on in the lesson. The seemingly ordinary conclusion reached at the end may be significant understanding for that pupil - that knowledge is subjectively constructed.

This descriptive entry is more useful than those in the previous term's experiment. It is more detailed and also summarizes what has been learned as well as what had been taught. Yet the potential of the log is not being used. It would have been valuable for that pupil to explore in writing why she chose overcrowding and others chose 'not having a bath'. What did she take into account in making her decision? Did the opinions of others influence her? Is the final sentence more, perhaps, an indication of not getting to grips with conflicting opinions?

The evaluative comments were also more developed than in term 1:

> I found today's lesson quite interesting because I like working with hypotheses and trying to find out if they are true or false. I am looking forward to writing the essay on land use in a city because I will be curious of how the city of Sheffield uses its land. I will be interested in the results.

This is a judgement - 'interesting' - plus an explanation of that judgement which reveals something about the student's attitudes and feelings. It enables the teacher to understand the student better, and also to understand how the activities he plans may be received.

205

Again the category with the smallest number of entries was reflection. Occasionally students did sort out for themselves what had been learned:

> At first glance the choropleth maps looked easy but they aren't that easy because suitable intervals have to be decided upon. The shading itself isn't as easy as it seemed because it is hard to do lots of different shapes.

In spite of these entries the teacher was still disappointed. The entries in the logs were very short and most of the writing was descriptive and evaluative. He was not certain that the pupils were clear about the purpose of the logs. Although the new instructions had helped define the purpose, his comments and replies, written after each entry, were giving a contradictory message. The comments were analysed and fell into the categories shown in table 11.1. All the comments were positively phrased and constructive, yet they were dominated by evaluation. The teacher made the students feel that they were being assessed as writing good or bad logs rather than entering into a dialogue with him. The problem was how to encourage more useful and valuable logs - which is, of course, making a judgement -without acting the role of judge. There had been no comments to engage the pupils' geographical minds. There were many missed opportunities. Almost all the pupils had given the teacher something to reply to, as the following examples show:

> I wasn't sure what the word ghetto implied.
>
> I could do with more background knowledge.

The teacher used the analysis and discussion to adjust his comments. It was also agreed during a useful discussion with pupils that from then on the logs would be voluntary. Only two pupils decided not to discontinue:

> I do see the point in writing these things. I thought they were useless ... The shy people, like me, have a lot to gain by this private form of communication.
>
> I decided to continue because it helps me to remember what we have learned in the lesson and you make an effort to listen more as you know you have to write about it in the learning log at the end. It will be interesting looking

Table 11.1 Comments in the logs

Category	Example	No. of comments in category
1 Evaluative		
(a) Positive	Some very interesting comments	9
(b) Negative	You need to develop more depth to your comments	14
2 Questions	Why was it difficult?	4
3 Expressing feelings	I'm glad you were satisfied	3
4 Replying to an entry	Talking to a group of people can be daunting	2
Total		32

back once you have finished the course and reading your thoughts and feelings on the work at the time.

These pupils were explicit about two important aspects of logs: as communication with the teacher, and as communication with oneself.

From that point on the teacher changed his comments radically. During the next six weeks they fell into the following categories:

1 Evaluative:
    (a)   Positive   2
    (b)   Negative   1
2 Questions   40
3 Feelings   8
4 Comment   20

The largest category of replies was 'questions'. These were probing understanding, seeking opinions, asking for reasons, debating the validity of exercises. Some of the questions were answered briefly in the logs, some were discussed orally, but they encouraged an opening up of the students' thinking. They seemed more aware of a listening, responsive audience. The students started challenging the teacher and asking questions:

> You didn't really explain why we are doing this and graphing it. Is it really important?
>
> Also didn't see the point in what we did. Those readings, can't have been half accurate. Wasn't worth doing all that for those kind of results.
>
> I managed to do the homework last night but I didn't enjoy it and it took me ages. (Approx. time taken 2-2 1/2 hours.)

The logs became more communicative from this point onward, and they were used by pupils to serve their purposes. They were more reflective, more argumentative, and were useful to both teacher and pupil.

The account of this experiment indicates that it is not enough merely to ask pupils to keep a diary. The role of the teacher in introducing the logs and in responding to them has an influence on what is written.

EXPERIMENT D

The third experiment with learning logs is still in progress. A sixth-form A-level group has been using them since the beginning of a Schools Council 16-19 course. This time, however, the logs have a totally different character which is illustrated in the extracts below:

> *Student.* The lessons are still very interesting but getting slightly harder.
>
> *Teacher.* Could you tell me a bit more about the 'hardness', explore that for me - do you mean you are uncertain about my instructions?
>
> *Student.* The lessons aren't as hard as all that now and I know how to do my homework for the next lesson. I think it is because I am getting more organized.
>
> *Teacher.* In what ways have you organized yourself?
>
> *Student.* I have organized myself by getting used to getting up early in a morning and going to bed earlier at nights. Also by working out what times I can do my homework for the next lesson. It makes the lesson easier when you know what people are talking about.

These extracts do not constitute the whole in either the pupil's entry or the teacher's reply but are selected to show a dialogue continuing over several entries. In experiment C the

Writing as reflection

teacher's questions were answered briefly, sometimes with just 'yes' or 'no'. In this example each question put by the teacher is picked up in the next entry. The continuity of discussion gives the student an opportunity to make herself clear to the teacher, to explain previous entries and to bring the teacher up to date: 'The lessons aren't as hard as all that now'! The type of question the teacher in experiment D is asking is rather different from that in C, where typical questions were asking for students' opinions on issues, students' judgements on activities, and pupils' suggestions for activities. The teacher in C is trying to understand the *course* through the students' perceptions. In contrast the purpose of the questions in experiment D is to understand the *students* rather than the course.

This attitude encourages more revelations.

> I did not completely understand it all. I got up to the scaled part of it, but then did not understand what was meant by the mean, either the total mean or the total scaled mean. I was confused.

The teacher's reply is supportive and invites further admissions of confusion:

> Do you realize that one of the first crucial steps required to deal with confusion is to actually identify where you are confused and then this can help you to focus on the problems and so overcome them?

If this student continues to identify problems and communicate them, then the writing and ensuing understanding between student and teacher will surely lead to improved learning.

Another distinctive feature of the logs is that the pupils are using them to understand the teacher:

> *Student.* I am a bit confused as to why we are doing two different pieces of work, one on cities and the other on evaluating mapping. I am not sure why they are tied together.
>
> *Teacher.* Mapping techniques affect your views of what reality is. In the past you have just skated over this - but now we need to explore it thoroughly ... The writing last night was exploratory for you - it's one way of dealing with uncertainties and exploring your own thoughts so you can learn. Rather than just

'absorbing' other people's ideas.

Student. I don't quite understand the above 'absorbing other people's ideas'. Whose ideas? The people in the group or people who write textbooks?

These are typical of the confusions and misunderstandings which take place in every teaching and learning situation but which are rarely voiced.

The logs in experiment D have enabled students and teacher to talk to each other regularly and privately. Both parties in the dialogue look forward to reading the other's entry. The logs are written once a week at the end of a double lesson, and the teacher types replies to each entry before the following lesson.

It is interesting to speculate about the elements in this experiment which have led to such fruitful dialogue. First, the group is small, just eight students, so the teacher feels she has time to reply to each every week. Second, the way the logs were introduced was different, partly by default. The teacher had mislaid written instructions and gave only brief oral instructions. She modelled the logs on 'letters to your teacher' which she had used in the USA. She emphasized that they were for anything the students wanted to say to her at all and that she was interested in their feelings. Her limited instructions meant that the students set the agenda for the logs. They are writing what they want to tell her, what they really want to communicate, not what they think she wants to know. In all the other experiments the teachers, by giving written instructions, have set the agenda. In experiment D, as the logs have developed the teacher has given no general explicit instruction to the whole group. For example, when pupils do not answer her questions she does not tell the whole group that she expects everyone to answer questions. This action would change the relationship she is building up. Instead she probes the individual:

> Have you read my questions after your last entry? I find that when you write a two or three-line entry I am left wondering what you mean.

The focus is on her understanding the student rather than the student completing a task. So the student, not wanting to be misunderstood, is encouraged to clarify what he means in the next entry. The emphasis in experiment D is on student-centred dialogue.

CONCLUSIONS

All the teachers/tutors involved in these experiments have found the entries in the logs, in their variety indicated above, valuable. The type of entry seems to depend on three variables. First, the diaries are written in a context of expectations about written work and about student-teacher relationships. They are written after specific classroom interactions. The context, including staff-student relationships, pedagogy, and expectations, is important. Reflective writing is encouraged by more equal classroom relationships and a pedagogy in which student participation is valued. Second, the instructions for logs can restrict what is written. The instructions may reflect the different interests of the teacher in the logs. They can be used primarily to inform the teacher about the course. Alternatively they can be a communication about the student. Third, teachers' responses to logs again reveal their own agendas for the logs and encourage different types of writing. Teachers cannot expect to be a neutral receiver of logs. Inevitably the logs are influenced by the context in which they are placed, by the instructions given, and by the nature of teacher-tutor response. In monitoring and adjusting these variables the teacher is also making decisions about whose purposes the logs are more likely to serve.

INDEX

Adam, A. 154-5
Armstrong, Michael 193
anti-racism 92-3
Apple, M.W. 163
Argyris, C. 119, 185, 186
Aschner, M.J. 75
Aspects of Secondary Education (DES) 11
audience, role of 12-14
Ayers, A. 155

Barnes, Douglas 2, 4, 6, 13-14, 75, 100, 193
Barrett, G. 135
Beddis, R. 167-8
bilingualism 95; and humanities writing 39-58
Bisset, N. 151
Bliss, J. 125, 132
Boardman, D. 72, 86-7
Boekaerts, M. 132
Bolwell, L. 157
Bradford, M.G. 165-6
Brandes, D. 179
Brice-Heath, S. 52
Britton, J. 2, 6, 43, 180
Bruner, Jerome S. 15, 24, 116-17
Burgess, T. 4
Burke, Edmund 153
Buzan, T. 123

Calvert, B. 178
Capel, H. 163
Champaigne, A. 132

class, and writing 55-7
classroom: basic functions 59-60; language 157-9; as writing environment 39-42, 66-9
Collaborative Learning Project 98, 104
comprehension, reading 24-5
concept development 4;
concept maps 115-37; writing about the environment 141-7
concept maps 115-37; and child 124-6; complexity 127-9; labelled-line 119-23; links between concepts 129-34; reconstruction of teaching 135; structures 116-19; and teacher 123-4
Creole 94
Crystal, D. 94
cultural pluralism 92
Cummings, R. 14

Dale, P.S. 96
D'Arcy, Pat 193
DARTS 18
diaries 175-91, 193-211; dialogue through 180-6; from teacher-centric to pupil-centric 178-9; learning about self and others 186-9; multiple realities 179-80; problems of 190-1; and video-

Index

recording 189-90
Dulay, H.C. 95
Duncan, S.S. 165
Dunford, M. 164
Dunlop, S. 154-5
Dunsbee, T. 97
Dyer, J. 132

Edwards, V. 93, 94
Elden, M. 191
English, M. 75
English, Standard 94
English as a Second Language (ESL) techniques 101-2, 105
Entwistle, N. 129, 178
environment, writing about 141-7
equality of opportunity 92
everyday life, writing about 141-7

Fairgrieve 164
Fear, J. and S. 102
Fenker, R. 132
Fien, J. 167
Fisher, S. 96, 110
Fontana, D. 52
Ford, T. 97
Foucault, M. 165
Frijda, N. 132

Gallagher, J.J. 75
Gardner, K. 16-18, 104-5, 107
geography: humanistic 167-9; ideology 151-60, 162-70; map work problem-solving 71-90; physical, case study 59-69; see also language and learning
George, N. 95
Ghaye, Anthony L. xi, 4, 5-6, 71, 115-37, 175-91
Giddens, A. 152, 162
Gilbert, Robert xi, 4-5, 151-60, 163, 166
Ginnis, P. 179
Goodman 24
Gray, F. 165
Gregory, D. 163-4, 167, 169-70

Habermas, J. 168
Halliday, M.K. 13
Hamilton-Wieler, Sharon xi, 2-3, 59-69
Harding, D.W. 13
Harvey, D. 162, 164
Hawkridge, D. 123
Held, D. 168-9
Henley, Richard xi, 5, 162-70
Hicks, D. 96, 110
Hodge, R. 151, 169
Holt, J. 190
Hopkins, D. 186
Houlton, D. 96
Hull, Robert 23
humanistic geography 167-9
humanities, writing 39-58

ideology 4-5; of geographical language 162-70; in geography teaching 151-60; and multicultural education 92-110; textbook language 154-7

Jacobs, Jane 142
Johnson, M. 156
Johnston, R.J. 164, 166
Jones, A. 95
Jones, Ann xi, 4, 141-7
Joyce, B. 123

Keat, R. 153
Kelly, George 24
Kent, W.A. 165-6
Kress, G. 50, 94, 151, 157, 169

Langbein 64

213

Index

language 163
Language across the Curriculum school 2-3, 11-37; learning process 21-3; and open approach to language development 102-3; reading and comprehension 24-5; Robson's study 25-37; textbooks 23-4; writing 19-21
language and learning 2-3, 11-37; audience 12-14; concept maps 115-37; diaries 175-91, 193-211; ideology of geographical language 162-70; ideology in geography teaching 151-60; mapwork 71-90; multicultural education 92-110; physical geography, case study 59-69; writing about the environment 141-7; writing, humanities classroom 39-58; writing as reflection 193-211
Lawson, A. 124
learning, process 21-3 *see also* language and learning
learning logs 193-211; on PGCE course 194-200; on sixth-form field trip 201-3; and syllabus evaluation 203-10; value of 196-200
Lee, R. 166
Leopold 64
Levine, J. 96
Lewis, B. 123
Lewis, Daniel xi, 2, 3, 39-58, 92-110
Lowenthal, D. 168
Lunnon, A.J. 71

Lunzer, E. 16-18, 104-5, 107
Lynch, K. 168

McHoul, A. 157
Macintosh, H.G. 72
McLeod, Alex 14-16
McLeod, Peter 59-67
map work, Reality Orientated Problem Solving (*q.v.*) 71-90
Marsden, W.E. 72, 163-4
Martin, Nancy 2, 6, 11, 13, 42
Marton, F. 121
Marx, Karl 163
Mehan, H. 176
Milburn, D. 23-4
Miliband, R. 164
Mills, C. 154
multicultural education 92-110; dialect, bilingualism and development of Standard English 93-6; geography curriculum 93; language and racism 108-10; open and structured approaches 102-8; perspectives on language 92-3; writing 55-7

Naidoo, Beverley 101
National Writing Project 42
Ness, T. 157
Nisbet, J. 137, 190
Novak, J. 123

Okpala, Julie I.N. xi, 3, 71-90
open approach to language development 97-8, 102-3
Owen, J.G. 72

Paris, S. 127
Parr, A.E. 147
Perrons, D. 164
physical geography, case study 59-69; A-level classrooms 59-60; 'certainty'

61-2; framing the question 60-1; 'I' and text 64-6; speculation 62-3; writing in the classroom 66-9; writing about fieldwork 63-4
Paiget, Jean 12-13, 71
Pocock, J. 153
Popper, Karl 124
positivism 165-7
Preusch, D. 102
problem-solving, reality orientated (*q.v.*) 71-90

race 55-7; *see also* multicultural education
racism 108-10; anti- 92-3
Raleigh, M. 96
reading, and comprehension 24-5
Reality Orientated Problem Solving (ROPS) 71-90; map work 72-3; misconceptions 76-88; productive verbalization 71-2; statement of problem 73-4
reflection and language 5-6; diaries 175-91; learning logs 193-211
Reiser, R. 168
Relph, E.L. 167-8
Renner, J. 124
Richmond, J. 42, 94, 104
Riley, S. 101
Roberts, Margaret xii, 6, 193-211
Robinson, Elizabeth G. xii, 4, 115-37
Robson, Carol 2, 3, 23-37
Role of Language in Learning movement *see* Language across the Curriculum
Rose, C. 168
Rosen, Harold 2, 6, 63

Rowntree, D. 123
Rudnitsky, A. 132, 135
Sarup, M. 167
Sayer, A. 163-4, 166
Schon, D. 119, 185, 186
Schwab, J. 116
Scott, S. 98, 104
Searle, C. 92
Shavelson, R. 132
Shucksmith, J. 137, 190
Slater, Frances xii, 1-6, 11-37, 104, 167
Smith, F. 24
Stewart, J. 132
Stone, W. 156
Stretton, Hugh 153
structured approach to language development 98-100
structures 116-19
Stuart, H. 132
Stubbs, M. 55
Sutton, S. 103

teacher: and concept maps 123-4; and diaries 178-86; and learning logs 193-211; and writing 45-6; *see also* language and learning
textbooks 23-4; ideology 154-7; and racism 108-10
Thompson, E.J. 156
Todd, Frankie 14
Towner, E. 86-7
Trudgill, P. 55, 94
Tuan 167-8

Upton, L. 127
Urry, J. 153

values education 110
Van Den Daele, L. 117
Vygotsky, L.S. 12-13, 15, 22-3, 81

Walford, R. 110, 159

215

Index

Ward, C. 147
Warren, W. 135
Watson, R. 157
Weil, M. 123
Williams, Raymond 147, 162
Woods, P. 186
Woolfenden, P. 123
writing: children compared to adults 50-1; classroom environment 39-42; diaries 175-91, 193-211; encouraging development 42; about the environment 141-7; about fieldwork 63-4; humanities 39-58; individual differences 51-3; Language across the Curriculum school 19-21; language background 53-5; learning logs 193-211; as open process 43-5; range of 43-4; as reflection 193-211; social factors 55-7; teacher attitudes 45-6
Writing across the Curriculum Project 42

Young, E. 156